英文運用

公務員招聘

英文運用

CRE解題王

有效提升 CRE 應考能力
一書掌握公務員英語程度

U0130815

- 由資深大專課程導師編撰

- 題目緊貼CRE形式及深淺

- 模擬試卷及答案詳盡講解

CRE專家
Fong Sir 著

序言

公務員薪高糧準，要成功通過公務員招聘，以學位／專業程度職系而言，最基本的要求就是通過公務員綜合招聘考試（Common Recruitment Examination-CRE），該測試首先包括三張各為45分鐘的多項選擇題試卷，分別是「中文運用」、「英文運用」、和「能力傾向測試」，其目的是評核考生的中、英語文能力及推理能力。

此外，基本法及香港國安法是一張設有中英文版本的選擇題形式試卷。全卷共20題，考生須於30分鐘內完成。考生如在20題中答對10題或以上，會被視為取得《基本法及香港國安法》測試的及格成績。

然而，一些考生雖有志加入公務員行列，但礙於此一門檻，因而未能加入公務員團隊。

有見及此，本書特為應考公務員綜合招聘試的考生提供試前準備，希望考生能熟習各種題型及答題方法。可是要在45分鐘之內完成全卷對大部分考生而言確有一定的難度。因此，答題的時間分配也是通過該試的關鍵之一。考生宜通過本書的模擬測試，了解自己的強弱所在，從而制訂最適合自己的考試策略。

此外，考生也應明白任何一種能力的培訓，固然不可能一蹴而就，所以宜多加推敲部分附有解說的答案，先從準確入手，再提升答題速度。考生如能善用本書，對於應付公務員綜合招聘考試有很大的幫助。

　　政府近年增加招聘人手，坊間湧現大量投考攻略，多以低價求售，但內容粗疏，對入職試認識不深。專家建議讀者選購時，應留意練習中是否包含「解題」，並選擇出版年資較深、口碑較佳的作者，便不會購入魚目混珠之作。

目錄

題庫練習一

I. Comprehension

This section aims to test candidates' ability to comprehend a written text. A prose passage of non-technical background is cited. Candidates are required to exercise skills in deciding on the gist, identifying main points, drawing inferences, distinguishing facts from opinion, interpreting figurative language, etc.

Exercise 1

There is extraordinary exposure in the United States to the risks of injury and death from motor vehicle accidents. More than 80 percent of all households own passenger cars or light trucks and each of these is driven an average of more than 11,000 miles each year. Almost one-half of fatally injured drivers have a blood alcohol concentration (BAC) of 0.1 percent or higher. For the average adult, over five ounces of 80 proof spirits would have to be consumed over a short period of time to attain these levels. A third of drivers who have been drinking, but fewer than 4 percent of all drivers, demonstrate these levels. Although less than 1 percent of drivers with BAC's of 0.1 percent or more are involved in fatal crashes, the probability of their involvement is 27 times higher than for those without alcohol in their blood.

There are a number of different approaches to reducing injuries in which intoxication plays a role. Based on the observation that excessive consumption correlates with the total alcohol consumption of a country's population, it has been suggested that higher taxes on alcohol would reduce both. While the heaviest drinkers would be taxed the most, anyone who drinks at all would be penalized by this approach.

To make drinking and driving a criminal offense is an approach directed only at intoxicated drivers. In some states, the law empowers police to request breath tests of drivers cited for any traffic offense and elevated BAC can be the basis for arrest. The National Highway Traffic Safety Administration estimates, however, that even with increased arrests, there are about 700 violations for every arrest. At this level there is little evidence that laws serve as deterrents to driving while intoxicated. In Britain, motor vehicle fatalities fell 25 percent immediately following implementation of the Road Safety Act in 1967. As Britishers increasingly recognized that they could drink and not be stopped, the effectiveness declined, although in the ensuing three years the fatality rate seldom reached that observed in the seven years prior to the Act.

Whether penalties for driving with a high BAC or excessive taxation on consumption of alcoholic beverages will deter the excessive drinker responsible for most fatalities is unclear. In part, the answer depends on the extent to which those with high BAC's involved in crashes are capable of controlling their intake in response to economic or penal threat. Therapeutic programs which range from individual and group counseling and psychotherapy to chemotherapy constitute another approach, but they have not diminished the proportion of accidents in which alcohol was a factor. In the few controlled trials that have been reported there is little evidence that rehabilitation

PART ONE
題庫練習
PART TWO
模擬試卷
PART THREE
考生急症室

programs for those repeatedly arrested for drunken behavior have reduced either the recidivism or crash rates. Thus far, there is no firm evidence that Alcohol Safety Action Project-supported programs, in which rehabilitation measures are requested by the court,have decreased recidivism or crash involvement for clients exposed to them, although knowledge and attitudes have improved. One thing is clear, however; unless we deal with automobile and highway safety and reduce accidents in which alcoholic intoxication plays a role, many will continue to die.

1. The author is primarily concerned with:

   A. interpreting the results of surveys on traffic fatalities

   B. reviewing the effectiveness of attempts to curb drunk driving

   C. suggesting reasons for the prevalence of drunk driving in the United States

   D. analyzing the causes of the large number of annual traffic fatalities

   E. making an international comparison of experience with drunk driving

2. It can be inferred that the 1967 Road Safety Act in Britain:

A. changed an existing law to lower the BAC level that defined driving while intoxicated

B. made it illegal to drive while intoxicated

C. increased the number of drunk driving arrests

D. placed a tax on the sale of alcoholic drinks

E. required drivers convicted under the law to undergo rehabilitation therapy

3. The author implies that a BAC of 0.1 percent:

A. is unreasonably high as a definition of intoxication for purposes of driving

B. penalizes the moderate drinker while allowing the heavy drinker to consume without limit

C. will operate as an effective deterrent to over 90 percent of the people who might drink and drive

D. is well below the BAC of most drivers who are involved in fatal collisions

E. proves that a driver has consumed five ounces of 80 proof spirits over a short time

PART ONE
題庫練習

PART TWO
模擬試卷

PART THREE
考生急症室

4. With which of the following statements about making driving while intoxicated a criminal offense versus increasing taxes on alcohol consumption would the author most likely agree?

A. Making driving while intoxicated a criminal offense is preferable to increased taxes on alcohol because the former is aimed only at those who abuse alcohol by driving while intoxicated.

B. Increased taxation on alcohol consumption is likely to be more effective in reducing traffic fatalities because taxation covers all consumers and not just those who drive.

C. Increased taxation on alcohol will constitute less of an interference with personal liberty because of the necessity of blood alcohol tests to determine BAC's in drivers suspected of intoxication.

D. Since neither increased taxation nor enforcement of criminal laws against drunk drivers is likely to have any significant impact, neither measure is warranted.

E. Because arrests of intoxicated drivers have proved to be expensive and administratively cumbersome, increased taxation on alcohol is the most promising means of reducing traffic fatalities.

5. The author cites the British example in order to:

   A. show that the problem of drunk driving is worse in Britain than in the U.S.

   B. prove that stricter enforcement of laws against intoxicated drivers would reduce traffic deaths

   C. prove that a slight increase in the number of arrests of intoxicated drivers will not deter drunk driving

   D. suggest that taxation of alcohol consumption may be more effective than criminal laws

   E. demonstrate the need to lower BAC levels in states that have laws against drunk driving

6. Which of the following, if true, most weakens the author's statement that the effectiveness of proposals to stop the intoxicated driver depends, in part, on the extent to which the high-BAC driver can control his intake?

   A. Even if the heavy drinker cannot control his intake, criminal laws against driving while intoxicated can deter him from driving while intoxicated.

   B. Rehabilitation programs aimed at drivers convicted of driving while intoxicated have not significantly reduced traffic fatalities.

   C. Many traffic fatalities are caused by factors unrelated to excessive alcohol consumption on the part

PART ONE
題庫練習

PART TWO
模擬試卷

PART THREE
考生急症室

of the driver.

D. Even though severe penalties may not deter intoxicated drivers, these laws will punish them for the harm they cause if they drive while intoxicated.

E. Some sort of therapy may be effective in helping problem drinkers to control their intake of alcohol, thereby keeping them off the road.

Answers 1:

1. B

This is a main idea question. The author begins by stating that a large number of auto traffic fatalities can be attributed to drivers who are intoxicated, and then reviews two approaches to controlling this problem: taxation and drunk driving laws. Neither is very successful. The author finally notes that therapy may be useful, though the extent of its value has not yet been proved. (B) fairly well describes this development.

(A) can be eliminated since any conclusions drawn by the author from studies on drunk driving are used for the larger objective described in (B).

(C) is incorrect since, aside from suggesting possible ways to reduce the extent of the problem, the author never treats the causes of drunk driving.

(D) is incorrect for the same reason.

Finally, (E) is incorrect, because the comparison between the U.S. and Britain is only a small part of the passage.

2. B

This is an inference question. In the third paragraph, the author discusses the effect of drunk driving laws, stating that after the implementation of the Road Safety Act in Britain, motor vehicle fatalities fell considerably. On this basis, we infer that the RSA was a law aimed at drunk driving. We can eliminate (D) and (E) on this ground.

(C) can be eliminated as not warranted on the basis of this information. It is not clear whether the number of arrests increased. Equally consistent with the passage is the conclusion that the number of arrests dropped because people were no longer driving while intoxicated. (C) is incorrect for a further reason, the justification for (B).

(B) and (A) are fairly close since both describe the RSA as a law aimed at drunk driving. But the last sentence of the third paragraph calls for (B) over (A).

As people learned that they would not get caught for drunk driving, the law became less effective. This suggests that the RSA made drunk driving illegal, not that it lowered the BAC required for conviction. This makes sense of the sentence " . . . they could drink and not be stopped." If (A) were correct, this sentence would have to read " . . . they could drink the same amount and not be convicted."

3. A

This is an inference question. In the first paragraph, the author states that to attain a BAC of 0.1 percent, a person would need to drink over five ounces of 80 proof spirits over a short period of time. The author is trying to impress on us that that is a consid-

PART ONE
題庫練習
PART TWO
模擬試卷
PART THREE
考生急症室

erable quantity of alcohol for most people to drink. (A) explains why the author makes this comment.

(B) is incorrect and confuses the first paragraph with the second paragraph.

(C) is incorrect since the point of the example is that the BAC is so high most people will not exceed it. This is not to say, however, that people will not drink and drive because of laws establishing maximum BAC levels. Rather, they can continue to drink and drive because the law allows them a considerable margin in the level of BAC.

(D) is a misreading of that first paragraph. Of all the very drunk drivers (BAC in excess of 0.1), only 1 percent are involved in accidents. But this does not say that most drivers involved in fatal collisions have BAC levels in excess of 0.1 percent, and that is what (D) says.

As for (E), the author never states that the only way to attain a BAC of 0.1 percent is to drink five ounces of 80 proof spirits in a short time—there may be other ways of becoming intoxicated.

4. A

This is an application question. In the second paragraph, the author states that increased taxation on alcohol would tax the heaviest drinkers most, but notes that this would also penalize the moderate and light drinker. In other words, the remedy is not sufficiently focused on the problem. Then, in the third paragraph, the author notes that drunk driving laws are aimed at the specific problem drivers. We can infer from this discussion that the author would likely advocate drunk driving laws over taxation for the reasons just given. This reasoning is presented in answer (A).

(B) is incorrect for the reasons just given and for the further reason that the passage never suggests that taxation is likely to be more effective in solving the problem. The author never really evaluates the effectiveness of taxation in reducing drunk driving.

(C) is incorrect for the reason given in support of (A) and for the further reason that the author never raises the issue of personal liberty in conjunction with the BAC test.

(D) can be eliminated because the author does not discount the effectiveness of anti-drunk driving measures entirely. Even the British example gives some support to the conclusion that such laws have an effect.

(E) is incorrect, for the author never mentions the expense or administrative feasibility of BAC tests.

5. C

This is a question about the logical structure of the passage. In paragraph 3, the author notes that stricter enforcement of laws against drunk driving may result in a few more arrests; but a few more arrests are not likely to have much impact on the problem because the number of arrests is small compared to those who do not get caught. As a consequence, people will continue to drink and drive. The author supports this with the British experience. Once people realize that the chances of being caught are relatively small, they will drink and drive. This is the conclusion of answer (C).

(A) is incorrect since the passage does not support the conclusion that the problem is any worse or any better in one country or the other.

(B) is incorrect since this is the conclusion the author is arguing against.

(D) is wrong because the author is not discussing the effectiveness of taxation in paragraph 3.

(E) is a statement the author would likely accept, but that is not the reason for introducing the British example. So choice (E) is true but nonresponsive.

6. A

This is an application question that asks us to examine the logical structure of the argument. In the fourth paragraph, the author argues that the effectiveness of deterrents to drunk driving will depend upon the drinker's ability to control consumption. But drunk driving has two aspects: drunk and driving. The author assumes that drunk driving is a function of drinking only as indicated by the suggestion that control of consumption is necessary as opposed to helpful.

(A) attacks this assumption by pointing out that it is possible to drink to excess without driving. It is possible that stiff penalties could be effective deterrents to drunk driving if not to drinking to excess.

(B) is incorrect because the author actually makes this point, so this choice does not weaken the argument.

(C) is incorrect since the author is concerned only with the problem of fatalities caused by drunk driving.

(D) can be eliminated since the author is concerned to eliminate fatalities caused by drunk driving. No position is taken on whether the drunk driver ought to be punished, only that the drunk driver ought to be deterred from driving while intoxicated.

(E) is not a strong attack on the argument since the author does leave open the question of the value of therapy in combating drunk driving.

PART ONE
題庫練習
PART TWO
模擬試卷
PART THREE
考生急症室

Exercise 2

The origin of the attempt to distinguish early from modern music and to establish the canons of performance practice for each lies in the eighteenth century. In the first half of that century, when Telemann and Bach ran the collegium musicum in Leipzig, Germany, they performed their own and other modern music. In the German universities of the early twentieth century, however, the reconstituted collegium musicum devoted itself to performing music from the centuries before the beginning of the "standard repertory," by which was understood music from before the time of Bach and Handel.

Alongside this modern collegium musicum, German musicologists developed the historical subdiscipline known as "performance practice," which included the deciphering of obsolete musical notation and its transcription into modern notation, the study of obsolete instruments, and—most importantly because all musical notation is incomplete - the re-establishment of lost oral traditions associated with those forgotten repertories. The cutoff date for this study was understood to be around 1750, the year of Bach's death. The reason for this demarcation was that the music of Bach, Handel, Telemann, and their contemporaries did call for obsolete instruments and voices and unannotated performing traditions. Furthermore, with a few exceptions, late baroque music had ceased to be performed for nearly a century, with the result that orally transmitted performing traditions associated with it were forgotten. In contrast, the notation in the music of Haydn and Mozart from the second half of the eighteenth century was more complete than in the earlier styles, and the instruments seemed familiar, so no "special" knowledge appeared necessary. Also, the music of Haydn and Mozart, having never ceased to be performed, had maintained some kind of oral tradition of performance practice.

1. It can be inferred that the "standard repertory" might have included music:

    A. that called for the use of obsolete instruments.

    B. of the early twentieth century.

    C. written by the performance practice composers.

    D. written before the time of Handel.

    E. composed before 1700.

2. According to the passage, performance practice in the early twentieth century involved all of the following EXCEPT:

    A. deciphering outdated music notation.

    B. studying instruments no longer in common use.

    C. reestablishing unannotated performing traditions.

    D. determining which musical instrument to use.

    E. transcribing older music into modern notation.

PART ONE
題庫練習
PART TWO
模擬試卷
PART THREE
考生急症室

3. According to the passage, German musicologists of the early twentieth century limited performance practice to pre-1750 works because:

A. special knowledge was generally not required to decipher pre-1750 music.

B. unannotated performing traditions had been maintained for later works.

C. generally speaking, only music written before 1750 had ceased to be performed.

D. the annotation for earlier works was generally less complete than for the works of Bach and Handel.

E. music written prior to 1750 was considered obsolete.

4. The author refers to performance practice as a "subdiscipline" probably because it:

A. was not sanctioned by the mainstream.

B. required more discipline than performing the standard repertory.

C. focused on particular aspects of the music being performed at the German universities.

D. involved deciphering obsolete musical notation.

E. involved performing the works that were being transcribed at the universities.

Answers 2:

1. A

It is reasonably inferable from the first paragraph as a whole that the "standard repertory" mentioned refers to the music of Bach and Telemann as well as to other ("modern") music from their time (first half of the eighteenth century).

In the second paragraph, the author mentions that the music of Bach, Telemann, and their contemporaries called for obsolete instruments. Thus, the standard repertory might have included music that called for the use of obsolete instruments, as choice (A) indicates.

2. D

Although the passage does indicate that early music often called for the use of obsolete instruments, the passage does not state explicitly that performance practice involved determining which musical instrument to use.

3. B

According to the passage, the German musicologists did not study the music of Mozart and Haydn (post-1750 music) because, among other reasons, their music, "having never ceased to be performed, had maintained some kind of oral tradition of performance practice". Unannotated music is music that is not written, but strictly oral. Choice (B) restates the author's point in these lines.

4. C

According to the passage, performance practice was developed alongside the modern (early twentieth-century) collegium musicum, which was part of the German university. While the modern collegium musicum performed music from before the time of Bach and Handel, scholars in the field of performance practice studied certain aspects (e.g., choice of instruments, deciphering notation) of music from the same time period.

PART ONE
題庫練習

PART TWO
模擬試卷

PART THREE
考生急症室

Exercise 3

The encounter that a portrait records is most tangibly the sitting itself, which may be brief or extended, collegial or confrontational. Renowned photographer Cartier-Bresson has expressed his passion for portrait photography by characterizing it as "a duel without rules, a delicate rape." Such metaphors contrast quite sharply with RichardAvedon's conception of a sitting. While Cartier-Bresson reveals himself as an interloper and opportunist, Avedon confesses—perhaps uncomfortably—to a role as diagnostician and (by implication) psychic healer: not as someone who necessarily transforms his subjects, but as someone who reveals their essential nature. Both photographers, however, agree that the fundamental dynamic in this process lies squarely in the hands of the artist.

A quite-different paradigm has its roots not in confrontation or consultation but in active collaboration between the artist and sitter. This very different kind of relationship was formulated most vividly by William Hazlitt in his essay entitled "On Sitting for One's Picture" (1823). To Hazlitt, the "bond of connection" between painter and sitter is most like the relationship between two lovers. Hazlitt fleshes out his thesis by recalling the career of Sir Joshua Reynolds. According to Hazlitt, Reynolds' sitters were meant to enjoy an atmosphere that was both comfortable for them and conducive to the enterprise of the portrait painter, who was simultaneously their host and their contractual employee.

1. The author of the passage quotes Cartier-Bresson in order to:

A. refute Avedon's conception of a portrait sitting.

B. provide one perspective of the portraiture encounter.

C. support the claim that portrait sittings are, more often than not, confrontational encounters.

D. show that a portraiture encounter can be either brief or extended.

E. distinguish a sitting for a photographic portrait from a sitting for a painted portrait.

2. Which of the following characterizations of the portraiture experience as viewed by Avedon is most readily inferable from the passage?

A. A collaboration

B. A mutual accommodation

C. A confrontation

D. An uncomfortable encounter

E. A consultation

PART ONE
題庫練習

PART TWO
模擬試卷

PART THREE
考生急症室

3. Which of the following best expresses the passage's main idea?

A. The success of a portrait depends largely on the relationship between artist and subject.

B. Portraits, more than most other art forms, provide insight into the artist's social relationships.

C. The social aspect of portraiture sitting plays an important part in the sitting's outcome.

D. Photographers and painters differ in their views regarding their role in portrait photography.

E. The paintings of Reynolds provide a record of his success in achieving a social bond with his subjects.

Answers 3:

1. B

The author of the passage quotes Cartier-Bresson in order to:

A. refute Avedon's conception of a portrait sitting. (No contrast is provided to refute Avedons, also Avedon was introduced later)

B. provide one perspective of the portraiture encounter. (Correct - Different views have been provided, one of them is of Cartier)

C. support the claim that portrait sittings are, more often than not, confrontational encounters. (No such claim is given by author)

D. show that a portraiture encounter can be either brief or extended. (No such thing is said)

E. distinguish a sitting for a photographic portrait from a sitting for a painted portrait. (No comparison is done bw painting or photography, all scenarios are of photographic)

2. E

Which of the following characterizations of the portraiture experience as viewed by Avedon is most readily inferable from the passage?

A. A collaboration

B. A mutual accommodation

C. A confrontation - this is said for CB

D. An uncomfortable encounter

E. A consultation (Correct - this can be inferred from the passage, "diagnostician and psychic healer")

3. C

Which of the following best expresses the passage's main idea?

A. The success of a portrait depends largely on the relationship between artist and subject. (Nothing is talked about success of portrait)

B. Portraits, more than most other art forms, provide insight into the artist's social relationships. (Nothing is talked about artist's social relationships)

C. The social aspect of portraiture sitting plays an important part in the sitting's outcome. (Nothing is talked about difference in sittings outcome)

D. Photographers and painters differ in their views regarding their role in portrait photography. (Correct- Passage talks about difference in sittings of photographic portralts and painting portraits)

E. The paintings of Reynolds provide a record of his success in achieving a social bond with his subjects (Does not cover the entire gist of the passage)

PART ONE
題庫練習

PART TWO
模擬試卷

PART THREE
考生急症室

Exercise 4

Historians sometimes forget that no matter how well they might come to know a particular historical figure, they are not free to claim a god-like knowledge of the figure or of the events surrounding the figure's life. Richard III, one of England's monarchs, is an apt case because we all think we "know" what he was like. In his play Richard III, Shakespeare provided a portrait of a monster of a man, twisted in both body and soul. Shakespeare's great artistry and vivid depiction of Richard has made us accept this creature for the man. We are prepared, therefore, to interpret all the events around him in such a way as to justify our opinion of him.

We accept that Richard executed his brother Clarence, even though the records of the time show that Richard pleaded for his brother's life. We assume that Richard supervised the death of King Henry VI, overlooking that there is no proof that Henry was actually murdered. And we recoil at Richard's murdering his two nephews, children of his brother's wife Elizabeth; yet we forget that Elizabeth had spent her time on the throne plotting to replace her husband's family in power with her own family. Once we appreciate the historical context, especially the actions of Richard's opponents, we no longer see his actions as monstrous. Richard becomes, if not lovable, at least understandable. What's more, when we account for the tone of the times during which Richard lived, as illuminated in literary works of that era such as Machiavelli's The Prince, Richard's actions seem to us all the more reasonable.

1. With which of the following statements would the author of the passage most likely agree?

   A. In Richard III, Shakespeare portrays the king as more noble than he actually was.

   B. The deeds of Elizabeth were even more evil than those of Richard III.

   C. Richard III may have been innocent of some of the crimes that Shakespeare leads us to believe he committed.

   D. Richard III may have had a justifiable reason for killing Henry VI.

   E. Shakespeare was unaware of many of the historical facts about the life of Richard III.

2. The author of the passage refers to Shakespeare's "great artistry and vivid depiction of Richard" most probably in order to:

   A. make the point that studying Richard III is the best way to understand Richard as a historical figure.

   B. explain why Richard III is widely acclaimed as one of Shakespeare's greatest works.

PART ONE
題庫練習
PART TWO
模擬試卷
PART THREE
考生急症室

C. contrast Shakespeare's depiction of Richard with how Richard might have described himself.

D. illustrate how historians might become prejudiced in their view of historical figures.

E. point out that historians should never rely on fictional works to understand and interpret historical events.

Answers 4:

1. C

With which of the following statements would the author of the passage most likely agree?

A. In Richard III, Shakespeare portrays the king as more noble than he actually was. (This is 180 answer - Shakespere portrayed him as a monster instead)

B. The deeds of Elizabeth were even more evil than those of Richard III. (Author referred Elizabeth but didn't compare her as more evil as such. He mentioned her so that he could show that one should not go against Richard just because somone portrayed him as a monster.)

C. Richard III may have been innocent of some of the crimes that Shakespeare leads us to believe he committed. (2nd paragraph is all in support of Richard - Correct)

D. Richard III may have had a justifiable reason for killing Henry VI. (He doesn't say he kills Henry VI - infact says there is no proof of his killing)

E. Shakespeare was unaware of many of the historical facts about the life of Richard III. (Never mentions this either. Says even if he was aware of his life, should not claim he has a godlike knowledge)

2. D

The author of the passage refers to Shakespeare's "great artistry and vivid depiction of Richard" most probably in order to:

A. make the point that studying Richard III is the best way to understand Richard as a historical figure.

B. explain why Richard III is widely acclaimed as one of Shakespeare's greatest works.

C. contrast Shakespeare's depiction of Richard with how Richard might have described himself.

D. illustrate how historians might become prejudiced in their view of historical figures. (Correct - artistry and vivid depiction is a taunt at Shakespere in how he shows Richard life from his point of view)

E. point out that historians should never rely on fictional works to understand and interpret historical events.

PART ONE
題庫練習

PART TWO
模擬試卷

PART THREE
考生急症室

Exercise 5

The phrase "alternative stable state" in ecology refers to the tendency of many ecosystems to have different, stable configurations of biotic and abiotic conditions across large time scales separated by what are called regime or phase shifts. Alternative stable state theory claims that instead of a forest transitioning slowly along a gradient toward a different stable state, that forest will reach a crucial tipping point (known as an ecological threshold) as deforestation occurs. Any change beyond this threshold will lead to a rapid change towards the second stable state of the biome in question, in this case a grassland.

Ecologists typically describe this theory with an analogy: Picture a set of three hills, between which are two valleys with a ball sitting in one of them that you want to push into the other. If you don't push enough, the ball just rolls back down to where it started, but if you give the ball a big enough push, it will roll all the way into the valley on the other side of the hill. From there, it would require a similarly big push to get the ball back to where it started. Stable states are the valleys – where the balls want to stay if no outside forces are involved. However, if there is a big enough change in the environment to cause the ball to roll all the way up the hill to its highest point (the ecological threshold), the ball can be forced from one valley into another relatively quickly – this is a phase shift.

Until recently, most work discussing alternative stable states was theoretical – the idea of purposefully changing an environment to this extent was considered unthinkable – but several cases of confirmed alternative stable states have been reported. For example, disrupting the balance of phosphorous in a clearwater lake can lead to out-of-control phytoplankton blooms. Reducing this addition of phosphorous has so far not been an effective way of stopping the phytoplankton blooms, leading researchers to think that the ecosystem has been pushed into a new local equilibrium. This idea of hysteresis – that the state of an environment depends at least in part on its history and not just its current state – is at the core of many debates surrounding alternative stable state theory. However, whether most ecosystems that can exist under multiple stable states will readily convert between the two is still an open debate.

1. Which one of the following scenarios is most analogous to the type of ecological change predicted by alternative stable state theory?

   A. After 30 years working at a bank, a man quits suddenly after a severe illness and then spends the rest of his life working as a patient advocate.

   B. The stock market slowly rises over a 10-year period and then crashes precipitously after unexpected news regarding the health of the economy.

- PART **ONE**
題庫練習

PART TWO
模擬試卷

PART THREE
考生急症室

C. Over a period of a few years, a company switches from manufacturing a particular software product to consulting on that product after sales slowed.

D. The price of a certain stock that has been constant for months spikes dramatically on rumors of a take-over bid but then quickly returns to the old price after the rumor is debunked.

E. The ecology of a lake slowly changes as silt deposits increase from frequent flooding, but the lake returns to its original state as floodwaters recede.

2. Which of the following is confirmed in the passage as causing a phase shift in the environment?

A. the addition of phosphorous to a lake

B. the blooming of phytoplankton in a lake

C. the movement of balls over a hill

D. the presence of drought conditions in a certain region

E. the repeated logging of certain forests

3. The primary purpose of the passage is to:

A. highlight the uncertainties and debates relating to a particular theory.

B. suggest that the alternative stable state theory is

incomplete.

C. argue that the alternative stable state theory is primarily correct.

D. provide a thorough understanding of an important ecological theory.

E. give several detailed examples of the alternative stable state theory.

PART ONE
題庫練習

PART TWO
模擬試卷

PART THREE
考生急症室

Answers 5:

1. A

The question is to find out which of the following options are most analogous to the stated theory. Well for these questions, first in your mind you chalk out a rough idea of what is the theory all about. Then use POE to get to the answer. In this passage, the alternative stable state" theory is basically when a RAPID change occurs to the ecological environment, the environment shifts to another stable state. The change happens RAPIDLY and the final state reached is STABLE. These are the two characteristics of the stated theory. Ok, lets go into the options.

A - quitting a job is instant. Instant means it is a RAPID change to lifestyle. I am liking this already. Ok then he spends the rest of his life as a patient advocate. This is STABILITY. I think we have got our answer.

B - Rises slowly, nope. Has to be rapid.

C - Over a period of few years, nope again has to be instantaneous.

D - Spikes is stock is rapid. Cool. However, it returns to its original. Alternate stable theory states it has to be a different stability. Out.

E - Slowly changes, wrong answer.

2. A

Lets first skim through the options and see what absolutely doesnt make sense. Then work our way from there.

B, C are absolutely not relevant, right? B says the blooming of phytoplankton in a lake. Think about it, does blooming of this cause a phase shift. Lol no! C says movement of balls. Haha this is even funnier. It was only mentioned to give an analogy. C is out as well. By the way, I do not mind getting silly options in my questions, the more silly the options, the easier our job to get to the right answer!

So A, D and E kind of makes sense. Now we have to delve into it a bit deeper.

A mentions addition of phosphorous. yes that is also mentioned in our passage - something on the lines of if there is a dis-balance of phosphorous, then phase shift might happen. Okay, this looks pretty good.

D might be true in real life, unfortunately this is not mentionedin the passage. Out.

E says logging. What does logging mean? is it deforestation? Because if it is deforestation then in the first paragraph it does mention deforestation causes phase shifts. Oh I know what logging is, it is basically felling trees and bunching them together and transported. Excessive logging MIGHT lead to deforestation, but well we do not have the information from the passage. This is definitely not an inference question, plus we have a strong option A.

3. D

When I go through the passage, I usually write down what the tone of the passage is. It helps especially with the main point questions. In my humble opinion, it feels like the tone of author is very neutral (=) and he is stating some facts. Now lets go through the options.

A - I would not say uncertainties. Out.

B - incomplete? no. there is an open debate about the subject, but it is in no way incomplete. And besides that is just the last line of the passage. Definitely cannot be the main idea of the passage.

C - author is not arguing, he is just stating facts.

D - ecological theory in question is "alternative stable state". Thorough understanding of the theory yes, it is thorough with analogies and examples. I am happy with this.

E - Author does give examples, but is that all? No this passage is not all about examples. Out.

PART ONE
題庫練習

PART TWO
模擬試卷

PART THREE
考生急症室

Exercise 6

The rich analyses of Fernand Braudel and his fellow Annales historians have made significant contributions to historical theory and research. In a departure from traditional historical approaches, the Annales historians assume (as do Marxists) that history cannot be limited to a simple recounting of conscious human actions, but must be understood in the context of forces that underlie human behavior. Braudel was the first Annales historian to gain widespread support for the idea that history should synthesize data from social sciences, especially economics, to provide a broader historical view of human societies over time (although Febvre and Bloch, founders of the Annales school, originated this approach).

Braudel conceived of history as the dynamic interaction of three temporalities. The first of these, the evenementielle, involved short-lived dramatic "events," such as battles, revolutions, and the actions of great men, which had preoccupied traditional historians like Carlyle. Conjonctures was Braudel's term for the larger, cyclical processes that might last up to half a century. The longue duree, a historical wave of great length, was for Braudel the most fascinating of the three temporalities. Here he focused on those aspects of everyday life that might remain relatively unchanged for centuries. What people ate, what they wore, their means and routes of travel—for Braudel these things create "structures" that define the limits of potential social change for hundreds of years at a time.

Braudel's concept of the longue duree extended the perspective of historical space as well as time. Until the Annales school, historians had taken the juridicial political unit—the the nation-state, duchy, or whatever—as their starting point. Yet, when such enormous time spans are considered, geographical features may have more significance for human populations than national borders. In his doctoral thesis, a seminal work on the Mediterranean during the reign of Philip II, Braudel treated the geohistory of the entire region as a "structure" that exerted myriad influences on human lifeways since the first settlements on the shores of the Mediterranean Sea.

And so the reader is given such arcane information as the list of products that came to Spanish shores from North Africa, the seasonal routes followed by Mediterranean sheep and their shepherds, and the cities where the best ship timber could be bought.

Braudel has been faulted for the imprecision of his approach. With his Rabelaisian delight in concrete detail, Braudel vastly extended the realm of relevant phenomena; but this very achievement made it difficult to delimit the boundaries of observation, a task necessary to beginning any social investigation. Further, Braudel and other Annales historians minimize the differences among the social sciences. Nevertheless, the many similarly designed studies aimed at both pro-

PART ONE
題庫練習
PART TWO
模擬試卷
PART THREE
考生急症室

fessional and popular audiences indicate that Braudel asked significant questions which traditional historians had overlooked.

1. The primary purpose of the passage is to:

   A. show how Braudel's work changed the conception
      of Mediterranean life held by previous historians.

   B. evaluate Braudel's criticisms of traditional and
      Marxist historiography

   C. contrast the perspective of the longue duree with
      the actions of major historical figures

   D. illustrate the relevance of Braudel's concepts to
      other social sciences

   E. outline some of Braudel's influential conceptions
      and distinguish them from conventional approach-
      es

2. The author refers to the work of Febvre and Bloch in
   order to:

   A. illustrate the limitations of the Annales tradition of
      historical investigation

   B. suggest the relevance of economics to historical in-
      vestigation

   C. debate the need for combining various sociological
      approaches

D. show that previous Annales historians anticipated Braudel's focus on economics

E. demonstrate that historical studies provide broad structures necessary for economic analysis

3. According to the passage, all of the following are aspects of Braudel's approach to history EXCEPT that he:

A. attempted to unify various social sciences

B. studied social and economic activities that occurred across national boundaries

C. pointed out the link between increased economic activity and the rise of nationalism

D. examined seemingly unexciting aspects of everyday life

E. visualized history as involving several different time frames

4. The passage suggests that, compared to traditional historians, Annales historians are:

A. more interested in other social sciences than in history

B. more critical of the achievements of famous historical figures

PART ONE
題庫練習

PART TWO
模擬試卷

PART THREE
考生急症室

C. more skeptical of the validity of most economic research

D. more interested in the underlying context of human behavior provided by social structure

E. more inclined to be dogmatic in their approach to history

5. The author is critical of Braudel's perspective for which of the following reasons?

A. It seeks structures that underlie all forms of social activity.

B. It assumes a greater similarity among the social sciences than actually exists.

C. It fails to consider the relationship between short-term events and long-term social activity.

D. It clearly defines boundaries for social analysis.

E. It attributes too much significance to conscious human actions.

6. The passage implies that Braudel would consider which of the following as exemplifying the longue durée?

I. The prominence of certain crops in the diet of a region

II. The annexation of a province by the victor in a war

III. A reduction in the population of an area following a disease epidemic

A. I only

B. III only

C. I and II only

D. II and III only

E. I, II, and III

7. Which of the following statements is most in keeping with the principles of Braudel's work as described in the passage?

A. All written history is the history of social elites.

B. The most important task of historians is to define the limits of potential social change.

C. Those who ignore history are doomed to repeat it.

D. People's historical actions are influenced by many factors that they may be unaware of.

E. History is too important to be left to historians.

PART ONE
題庫練習
PART TWO
模擬試卷
PART THREE
考生急症室

Answers 6:

1. E

Primary purpose of the passage can be ascertained from the tone of the passage. The tone of the author here is describing someone's work and even justifying the merits of it. It is also describing how it differs from the conventional approaches. E is the closest choice.

Let us however see what the other options are.

A - too specific

B - wait what? surely not. Out.

C - historical figures vs longue duree? Either I have not understood the passage at all or this choice is definitely wrong!

D - too specific

2. D

I am not going to go into the details of why the others are wrong. It is pretty straight forward that it tells you Braudel wasn't the first, which means other historians anticipated it before. A, B, C, E are completely missing the point.

3. C

A, B, D, and E are all mentioned in the passage. The answer is arrived at by POE.

4. D

Well in this answer C and D are close choices. lets go through each.

C - oh wait, they advocate the use of economic data to further their historical research. This is out then.

D - In the paragraph - "but must be understood in the context of forces that underlie human behavior." Cool so this is our answer.

5. B

It is stated in the passage - "Further, Braudel and other Annales historians minimize the differences among the social sciences." So must be choice B.

6. A

Quick pre-thinking of what longue duree is and what will braudel suggest. Longue duree is little-little things over vast time-frames which never changes. Also it is not dependent on national boundaries but geographical boundaries. Ok so now lets go through each of the sentences.

I - A region is a geographical location. Diet of the region will remain unchanged (due to climate/location etc.) Ok this Braudel will agree with Keep this in.

II - Nope, not dependent on national boundaries. Must be larger.

III - A disease is going to last only a few years if that. To use the words of the passage, this would be classified as a evenementielle temporality. Think I am enjoying these RCs far too much!

Only I is correct then. Answer A.

7. E

We should ideally reach POE to reach to the answer. The other choices are quite weak. E is our answer because in the passage it is stated - "In a departure from traditional historical approaches, the Annales historians assume (as do Marxists) that history cannot be limited to a simple recounting of conscious human actions, but must be understood in the context of forces that underlie human behavior." The underlined part is what traditional historians do, hence it can be inferred that history is too important to be left to historians.

PART ONE
題庫練習

PART TWO
模擬試卷

PART THREE
考生急症室

Exercise 7

Isolated from the rest of the world by circumpolar currents and ice fields, the Dry Valleys of Antarctica never see snowfall. They are the coldest, driest places on earth, yet this arid, frigid climate supports several delicate ecosystems. Life in these ecosystems consists of relatively few groups of algae, microorganisms, and invertebrates, as well as plants such as lichen and fungi that live beneath the surface of rocks, where just enough light penetrates for photosynthesis to occur during the short period each year when meltwater is available. This region has proven interesting specifically to scientists researching the possibility of life on Mars because the features of the Dry Valleys are strikingly similar to the Martian landscape.

The Dry Valleys' system of lakes provides a particularly interesting area of research. Lake Hoare boasts a clear ice sun-cover fifteen meters thick that intensifies solar radiation in a way similar to solar panels; summer temperatures in bottom waters can become as high as 25° C, solely from solar heating, and not geothermal heating. This temperature permits huge mats of cyanobacteria to survive and even thrive on the lake floor. Pieces of this mat occasionally break free, floating up to the underside of the icy lake cover to melt through the ice toward the surface. The intense Antarctic winds, sweeping across the lake's surface, cause the ice to sublimate—turn directly from ice into vapor—rapidly. Such continuous sublimation should eventually cause the lake to vanish, but a continuous trickle of water from a nearby glacier — melted by the sun — refreshes the water. With a constant source of water, however small, and the heat generated by the solar radiation and retained by insulation from the thick ice cover, this lake offers an odd paradox: thick ice is responsible for maintaining this lake's liquid state.

The ancient river deltas around the slopes of Lake Fryxell present one feature typical of all river deltas: sediment thick enough to bury fossil life forms. This very thickness, however, makes a formal scientific search for signs of - life impracticable. The floor of Lake Vida, on the other hand, is covered with discrete piles of sediment up to a meter high that have preserved clear signs of life because rocks lying on the lake's surface, heated by sunlight, melt their way through the thick ice cover. Since smaller rocks, with a larger surface-area-tovolume ratio, get warmer and sink lower than larger rocks, pieces of gravel penetrate by as much as a meter, forming cracks in the ice that cause the finest sediment to sink even deeper. When the lake dries seasonally, the refined sediment within the ice drops to the lake floor, leaving a protective layer of gravel on top of the finer sediment. Dried bacteria in Lake Vida sediment have been dated back tens of thousands of years. Some researchers are hoping that exploration of similar terrain on Mars may yield similar results.

1. According to the passage, thick sediment found in the ancient river deltas of Lake Fryxell:

    A. forms layers of deposits laden with dried bacteria that are tens of thousands of years old

    B. masks life forms by continuously depositing new layers on top of older ones

    C. makes it unlikely that scientists will search there for traces of life

    D. collects liquid water from nearby glaciers warmed by sunlight during the Antarctic summer

    E. is devoid of life due to the impenetrable barrier formed by the sediment

2. The passage is primarily concerned with:

    A. the adaptations of microorganisms that allow them to live in the Dry Valleys' hostile environment

    B. the relationship between frozen lakes and glaciers which contributes to the availability of fresh water in the Dry Valleys

    C. evidence of past and present life forms in the extreme conditions of Antarctica's Dry Valleys

    D. the evolutionary histories of ancient lakes and the clues they hold about life in cold, dry ecosystems

    E. the differences and similarities between ancient river deltas and ancient lakes

3. Based on the information in the passage, scientists looking for life on Mars in conditions similar to those in the
Dry Valleys would be most likely to find it in which of
the following areas?

A. former river deltas

B. circumpolar ice fields

C. larger rocks

D. former glaciers

E. former lakebeds

4. Based on the information in the passage, mounds of
sediment found at the bottom of Lake Vida are refined
by:

I. Continual sublimation and ice cover

II. Large pieces of rock decomposing on the lake floor

III. Cracks formed in the ice by sinking rocks and gravel

A. I only

B. I and II

C. III only

D. II and III

E. I, II, and III

PART ONE
題庫練習

PART TWO
模擬試卷

PART THREE
考生急症室

Answers 7:

1. C

I am still confused b/w B & C, but here is my reasoning.

A. forms layers of deposits laden with dried bacteria that are tens of thousands of years old (Incorrect - since this happens in lake Hoare, as mentioned in second para from 2nd line, so this option is out.)

B. masks life forms by continuously depositing new layers on top of older ones. (Incorrect- its not mentioned anywhere in the passage for Lake Fryxell.)

C. makes it unlikely that scientists will search there for traces of life (Correct - as its mentioned in para 3 that its makes the search of scientists impracticable, meaning: it its unlikely that scientists would look for life there.)

D. collects liquid water from nearby glaciers warmed by sunlight during the Antarctic summer (Incorrect - since this is the case with lake Hoare, as mentioned in para 2.)

E. is devoid of life due to the impenetrable barrier formed by the sediment. (Incorrect- as its mentioned that this thick enough to bury all fossil life forms, implying that fossils are still present, implying that it is not completely devoid of life.)

2. C

A. the adaptations of microorganisms that allow them to live in the Dry Valleys' hostile environment (Incorrect - The purpose of the passage is clearly mentioned at the end of the 1st para.)

B. the relationship between frozen lakes and glaciers (Incorrect - unrelated to the purpose.)

C. evidence of past and present life forms in the extreme conditions of A which contributes to the availability of fresh water in the Dry Valleys (Correct - since the purpose of the passage is to find evidence of life in Dry valleys, since they are have the most similar conditions to Mars.)

D. the evolutionary histories of ancient lakes and the clues they hold about life in cold, dry ecosystems (Incorrect - this might be the secondary purpose, or the one that we need to compare the terrain of Mars with, but not a direct purpose.)

E. the differences and similarities between ancient river deltas and ancient lakes (Incorrect - unrelated to the purpose.)

3. E

A. former river deltas - incorrect, but close one, if we overlook it, this might seem the right answer, but when we look closely, we find that this is incorrect.

B. circumpolar ice fields (Incorrect)

C. larger rocks (Incorrect)

D. former glaciers (Incorrect, as not mentioned in the passage)

E. former lakebeds (Correct- as mentioned in the last para, regarding the lake beds of Lake Vida, Dried bacteria in Lake Vida sediment have been dated back tens of thousands of years. Some researchers are hoping that exploration of similar terrain on Mars may yield similar results.)

4. E

I, II, and III - Correct, since all the options are correct

Exercise 8

History has shaped academic medical centers (AMCs) to perform 3 functions: patient care, research, and teaching. These 3 missions are now fraught with problems because the attempt to combine them has led to such inefficiencies as duplication of activities and personnel, inpatient procedures that could and should have been outpatient procedures, and unwieldy administrative bureaucracies.

One source of inefficiency derives from mixed lines of authority. Clinical chiefs and practitioners in AMCs are typically responsible to the hospital for practice issues but to the medical school for promotion, marketing, membership in a faculty practice plan, and educational accreditation. Community physicians with privileges at a university hospital add more complications. They have no official affiliation with the AMC's medical school connected, but their cooperation with faculty members is essential for proper patient treatment. The fragmented accountability is heightened by the fact that 3 different groups often vie for the loyalty of physicians who receive research. The medical school may wish to capitalize on the research for its educational value to students; the hospital may desire the state-of-the-art treatment methods resulting from the research; and the grant administrators may focus on the researchers' humanitarian motives. Communication among these groups is rarely coordinated, and the physicians may serve whichever group promises the best perks and ignore the rest—which inevitably strains relationships.

Another source of inefficiency is the fact that physicians have obligations to many different groups: patients, students, faculty members, referring physicians, third-party payers, and staff members, all of whom have varied expectations. Satisfying the interests of one group may alienate others. Patient care provides a common example. For the benefit of medical students, physicians may order too many tests, prolong patient visits, or encourage experimental studies of a patient. If AMC faculty physicians were more aware of how much treatments of specific illnesses cost, and of how other institutions treat patient conditions, they would be better practitioners, and the educational and clinical care missions of AMCs would both be better served.

A bias toward specialization adds yet more inefficiency. AMCs are viewed as institutions serving the gravest cases in need of the most advanced treatments. The high number of specialty residents and the presence of burn units, blood banks, and transplant centers validate this belief. Also present at AMCs, though less conspicuous, are facilities for ordinary primary care patients. In fact, many patients choose to visit an AMC for primary care because they realize that any necessary follow—up can occur almost instantaneously. While AMCs have emphasized cutting-edge specialty medicine, their more routine medical services need development and enhancement.

PART ONE
題庫練習
PART TWO
模擬試卷
PART THREE
考生急症室

A final contribution to inefficiency is organizational complacency. Until recently, most academic medical centers drew the public merely by existing. The rising presence, however, of tertiary hospitals with patient care as their only goal has immersed AMCs in a very competitive market. It is only in the past several years that AMCs have started to recognize and develop strategies to address competition.

1. The author's attitude toward the inefficiencies at academic medical centers is one of:

    A. reluctant acquiescence

    B. strident opposition

    C. agonized indecision

    D. reasoned criticism

    E. enthusiastic support

2. The author of the passage would most likely agree with which of the following statements about primary care at AMCs?

    A. AMCs would make more money if they focused mainly on primary care.

    B. Burn and transplant patients need specialty care more than primary care.

    C. AMCs offer the best primary care for most patients.

D. AMCs have not tried hard enough to publicize their primary care services.

E. Inefficiencies at AMCs would be reduced if better primary care were offered.

3. The author's primary purpose in this passage is to:

A. discuss the rise and fall of academic medical centers

B. explain that multiple lines of authority in a medical center create inefficiencies

C. delineate conflicts occurring in academic medical facilities

D. examine the differences between academic and other health care entities

E. warn that mixed accountabilities result in treatment errors

4. The author implies which of the following about faculty physicians at AMCS?

A. Most of them lack goodlbusiness sense.

B. They put patients' physical health above their hospitals' monetary concerns.

C. They sometimes focus on education at the expense of patient care.

PART ONE
題庫練習

PART TWO
模擬試卷

PART THREE
考生急症室

D. They lack official affiliation with the medical schools connected to AMCS.

E. They choose AMCS because follow—up care can be given very quickly.

5. Which of the following would the author probably consider a good strategy for academic medical centers dealing with competition from tertiary hospitals?

A. recruiting physicians away from tertiary centers

B. increasing the focus on patient care

C. sending patients to tertiary facilities

D. eliminating specialty care

E. reducing dependence on grant money

Answers 8:

1. D

The author's attitude toward the inefficiencies at academic medical centers is one of

A. reluctant acquiescence (Not correct: the author of the passage mentions the reasons for the inefficiencies in AMCs. Passage doesn't convey that the author reluctantly accepts any of those inefficiencies)

B. strident opposition (Not correct: the author just mentions the reasons for the inefficiencies, no opposition to those inefficiencies)

C. agonized indecision (Not correct: the passage doesn't say that the author is pained by these inefficiencies)

D. reasoned criticism (Correct: in the passage, the author gives clear reasons for the inefficiencies, and by saying 'organisational complacency' in the last para, he displays criticism)

E. enthusiastic support (Not correct: No where in the passage, author supports these inefficiencies)

2. B

The author of the passage would most likely agree with which of the following statements about primary care at AMCs?

A. AMCs would make more money if they focused mainly on primary care. (Not correct: discussion on money is not given in the passage)

B. Burn and transplant patients need specialty care more than primary care. (Correct: passage says 'AMCs are viewed as institutions serving the gravest cases in need of the most advanced treatments. The high number of specialty residents and the presence of burn units, blood banks, and transplant centers validate this belief' and considers primary care as ordinary care)

C. AMCs offer the best primary care for most patients. (Not correct: no comparison between AMCs and other medical centres is given in the passage)

D. AMCs have not tried hard enough to publicize their primary care services. (Not correct: passage says AMCs drew public merely by their existence, but no info is given to suggest that AMCs have not tried publicising their services)

E. Inefficiencies at AMCs would be reduced if better primary care were offered. (Not correct: many reasons for inefficiencies are listed in the passage, but how to reduce them is not discussed)

3. B

The author's primary purpose in this passage is to:

A. discuss the rise and fall of academic medical centers (Not correct: passage discusses mainly about the reasons for inefficiencies in AMCs, not about their rise and fall)

B. explain that multiple lines of authority in a medical center create inefficiencies (Correct: the passage mainly discusses about inefficiencies in AMCs and the reasons, which are mostly associated with authority & responsibilities)

C. delineate conflicts occurring i.n academic medical facilities (Not correct: the passage doesn't describe conflicts)

D. examine the differences between academic and other health care entities (Not correct: no comparison between academic and other health care entities)

E. warn that mixed accountabilities result in treatment errors (Not correct: treatment errors are not discussed)

4. C

The author implies which of the following about faculty physicians at AMCS?

A. Most of them lack goodlbusiness sense. (Not correct: no mention of physicians lacking business sense)

B. They put patients' physical health above their hospitals' monetary concerns: (Not correct: no comparison between patients' health and hospitals' monetary concerns)

C. They sometimes focus on educa-tion at the expense of patient care. (Correct: passage says 'Satisfying the interests of one group may alienate others')

D. They lack official affiliation with the medical schools connected to AMCS. (Not correct: no mention of physicians' affiliation with med schools)

E. They choose AMCS because follow—up care can be given very quickly. (Not correct: no info is given to suggest that physicians choose AMCs because of quick follow-up care)

5. B

A. recruiting physicians away from tertiary centers (Not correct: recruiting physicians away from tertiary centres will not change their goals, so competition will not change)

B. increasing the focus on patient care (Correct: if AMCs focus more on patient care, then as per last para, they can handle the competition)

C. sending patients to tertiary facilities (Not correct: not discussed in the last para)

D. eliminating specialty care (Not correct: not mentioned in the last para as a strategy to handle the competition)

E. reducing dependence on grant money (Not correct: not mentioned in in the last para as a strategy to handle the competition)

II. Error Identification

Knowledge on use of the language is tested through identification of language errors which may be lexical, grammatical or stylistic.

PART ONE
題庫練習
PART TWO
模擬試卷
PART THREE
考生急症室

Example :

The sentence below may contain a language error. Identify the part (underlined and lettered) that contains the error or choose 'E No error' where the sentence does not contain an error.

Irrespective for the outcome of the probe, the whole **sorry** affair has already **cast** a shadow over this man's **hitherto** unblemished record as a loyal servant to his country.

A. Irrespective for

B. sorry

C. cast

D. hitherto

E. No error

Answer : A

Questions:

1. He stopped <u>to smoke</u> because cigarettes <u>are</u> harmful <u>to</u> his <u>health</u>.

    A. to smoke
    B. are
    C. to
    D. health
    E. No Error

2. John <u>took off</u> his shoes so as <u>to not</u> <u>make</u> <u>any</u> noise.
    A. took off
    B. to not
    C. make
    D. any
    E. No Error

3. My <u>younger</u> brother <u>has</u> worked in <u>a</u> bank <u>since</u> a long time.
    A. younger
    B. has
    C. a
    D. since
    E. No Error

PART ONE
題庫練習

PART TWO
模擬試卷

PART THREE
考生急症室

4. These televisions <u>are</u> <u>too expensive</u> for <u>we</u> to <u>buy</u> at this time.

A. are

B. too expensive

C. we

D. buy

E. No Error

5. He was <u>angrily</u> <u>when</u> he <u>saw</u> what <u>was happening</u>.

A. angrily

B. when

C. saw

D. was happening

E. No Error

6. I <u>go</u> to Mexico <u>with</u> my girlfriend in <u>the</u> summer <u>of</u> 2006.

A. go

B. with

C. the

D. of

E. No Error

7. <u>That</u> is <u>the</u> man <u>which</u> told me <u>the</u> bad news.

   A. That

   B. the

   C. which

   D. the

   E. No Error

8. The books <u>writing by</u> Mark Twain are <u>very</u> popular <u>in</u> the world.

   A. writing

   B. by

   C. very

   D. in

   E. No Error

9. Nobody in this class <u>know</u> <u>anything about</u> the <u>changes</u> in the <u>exam formats</u>.

   A. know

   B. anything about

   C. changes

   D. exam formats

   E. No Error

PART ONE
題庫練習

PART TWO
模擬試卷
PART THREE
考生急症室

10. I <u>find</u> him absolutely <u>fascinated</u> when <u>talking</u> <u>with</u> him.

    A. find

    B. fascinated

    C. talking

    D. with

    E. No Error

11. When I arrived <u>at</u> our school, all the students <u>took</u> part <u>in</u> the musical show <u>had been</u> there.

    A. at

    B. took

    C. in

    D. had been

    E. No Error

12. Her <u>last</u> book <u>is</u> <u>published</u> in 20 languages <u>years</u> ago.

    A. last

    B. years

    C. is

    D. published

    E. No Error

13. The primary causes of species extinction or endanger-
    ment are habitat destruction, commerce exploitation
    and pollution.

    A. The primary causes

    B. are

    C. commerce

    D. pollution

    E. No Error

14. The books writing by Mark Twain are very popular in
    the world.

    A. writing

    B. in

    C. very

    D. by

    E. No Error

15. The exercises are so difficult for us to finish in one hour.

    A. The exercises are

    B. so difficult

    C. us to

    D. in one hour

    E. No Error

PART ONE
題庫練習
PART TWO
模擬試卷
PART THREE
考生急症室

16. About <u>two-third</u> <u>of my students</u> <u>wish</u> <u>to get</u> a scholarship to study abroad.

    A. two-third

    B. of my students

    C. wish

    D. to get

    E. No Error

17. Please help me <u>tidy</u> the room by <u>throwing</u> these <u>invaluable</u> things <u>away</u>.

    A. tidy

    B. throwing

    C. invaluable

    D. away

    E. No Error

18. The Call of the Wild is <u>one of</u> the <u>many</u> stories <u>about</u> the gold rush in Alaska <u>writing</u> by Jack London.

    A. one of

    B. many

    C. about

    D. writing

    E. No Error

19. There's <u>the woman</u> who <u>she</u> sold <u>me</u> the <u>handbag</u>.

    A. handbag

    B. the woman

    C. me

    D. she

    E. No Error

20. We were <u>advised</u> not <u>drinking</u> the <u>water</u> in <u>the</u> bottle.

    A. water

    B. advised

    C. the

    D. drinking

    E. No Error

21. I can't stand <u>make</u> noise in class. Would you please <u>do</u> <u>something</u> more <u>useful</u> ?

    A. something

    B. make

    C. useful

    D. do

    E. No Error

PART ONE
題庫練習
PART TWO
模擬試卷
PART THREE
考生急症室

22. If you listen <u>to</u> the questions <u>carefully</u>, you <u>would answer</u> them <u>easily</u>.

    A. to

    B. carefully

    C. would answer

    D. easily

    E. No Error

23. <u>Different</u> conversation efforts <u>have been made</u> in order to <u>saving</u> <u>endangered species</u>.

    A. different

    B. endangered species

    C. saving

    D. have been made

    E. No Error

24. The fire spread <u>through</u> the building very <u>quick</u> and the firemen failed <u>to put</u> it <u>out</u>.

    A. through

    B. quick

    C. to put

    D. out

    E. No Error

25. Our class <u>will visit</u> <u>the</u> teacher when we <u>will</u> have <u>free time</u>.

    A. will visit

    B. the

    C. will

    D. free time

    E. No Error

26. If we have a chance <u>to travel</u> abroad, Paris <u>is</u> the first city <u>where</u> we'd like <u>to visit.</u>

    A. to travel

    B. is

    C. where

    D. to visit

    E. No Error

27. They <u>have been asked</u> me <u>to visit</u> them <u>for ages</u>, but I <u>have never had</u> the time.

    A. have been asked

    B. to visit

    C. for ages

    D. have never had

    E. No Error

PART ONE
題庫練習

PART TWO
模擬試卷

PART THREE
考生急症室

28. If <u>I</u> were you, I <u>would have forgot</u> about <u>buying</u> a <u>new car</u>.

    A. I

    B. would have forgot

    C. buying

    D. new car

    E. No Error

29. People <u>said</u> that Henry <u>was</u> the <u>last</u> person <u>leaving</u> the burning house.

    A. said

    B. was

    C. last

    D. leaving

    E. No Error

30. The <u>equipment</u> in the office <u>was</u> <u>badly</u> in need of <u>to be repaired</u>.

    A. equipment

    B. was

    C. badly

    D. to be repaired

    E. No Error

31. There <u>are</u> probably <u>around</u> 3000 languages <u>speaking</u> <u>in</u> the world.

    A. are

    B. around

    C. speaking

    D. in

    E. No Error

32. The purpose <u>of</u> volunteer work <u>is</u> to <u>help</u> poor people how <u>improving</u> their life.

    A. of

    B. is

    C. help

    D. improving

    E. No Error

33. An <u>increasing</u> number of companies <u>has changed</u> dress codes, allowing employees <u>to wear</u> casual clothing <u>in the work place</u>.

    A. increasing

    B. has changed

    C. to wear

PART ONE
題庫練習

PART TWO
模擬試卷

PART THREE
考生急症室

D. in the work place

E. No Error

34. The woman who daughter I saw yesterday is a doctor.

    A. The woman

    B. who

    C. saw

    D. is

    E. No Error

35. Living in the city is more comfortably than in the country.

    A. in the city

    B. more

    C. comfortably

    D. in the country

    E. No Error

36. The aim of <u>this exercises</u> is <u>to practise</u> using will <u>for</u> a future fact or <u>prediction</u>.

    A. this exercises
    B. to practise
    C. for
    D. prediction
    E. No Error

37. We were <u>advised</u> not <u>drinking</u> <u>the water</u> <u>in</u> the bottle.

    A. advised
    B. drinking
    C. the water
    D. in
    E. No Error

38. <u>Having achieved</u> <u>highest score</u>, our team <u>was rewarded</u> <u>the first</u> prize.

    A. Having achieved
    B. highest score
    C. was rewarded
    D. the first
    E. No Error

PART ONE
題庫練習
PART TWO
模擬試卷
PART THREE
考生急症室

39. Ralph <u>wishes that</u> he went to the bank this morning <u>before</u> he <u>went</u> <u>to work</u>.

    A. wishes that

    B. before

    C. went

    D. to work

    E. No Error

40. The lesson was <u>too</u> difficult <u>for me</u> <u>to understand</u> <u>it</u>.

    A. too

    B. for me

    C. to understand

    D. it

    E. No Error

Answers:

1. A

- Stop doing something: no longer do something

- Stop to do something means stop in order to do other thing

2. B

- To not must be ed into "to not" to comply the pattern: so as not to do something

3. D

- Since + a specific time

- For + a period of time

- since must be ed into for

4. C

- we must be ed to be "us" to comply the rule of object going after preposition

5. A

- Angrily → angry (adjective)

6. A

- go must be ed into went

- Put verb in simple past tense to tell about an action happened in specific time in the past

7. C

- Who is used in relative clause for human, which for things

- Which → who

8. A

- writing must be ed into "written" to form a past participle phrase to say about a passive action 'books written by..."

9. A

- no one, nobody are always singular and, therefore, require singular verb

- know must be ed to be "knows"

10. B

- fascinated must be changed into "fascinating"

11. B

- Took → had taken

12. C

- Is → was

13. C

- Commerce → commercial

PART ONE
題庫練習

PART TWO
模擬試卷

PART THREE
考生急症室

14. A

- writing must be ed to be 'written" to form a past participle phrase to say about a passive action 'the books written by...

15. B

- so difficult → too difficult

16. A

- Two-third → two-thirds

17. C

- Invaluable → valueless (no value)

18. D

- Writing → written

19. D

- She must be omitted to form relative clause with who being subject of the relative clause

20. D

- Drinking → to drink (be advised to do something)

21. B

- Make → making (can't stand doing something)

22. C

- would answer → will answer

- Conditional sentence of type 1: If Clause in simple present, Clause in simple future

23. C

- Saving → save

24. B

- Quick → quickly

25. C

- will must be omitted to follow the pattern: subject will do when subject do / does to express actions in future

26. C

- Where → which

27. A

- Have been asked → have asked

28. B

- Conditional sentence of type 2: If subject did / were, ... would do....is used to say about untrue situation at present

- Would have forgot → would forget

29. D

- Leaving → to leave

30. D

- To be repaired → repair

31. C

- Speaking → spoken (passive meaning)

32. D

- Improving → to improve

33. B

- Has changed → have changed

34. B

- who must be ed to be "whose" as a possessive pronoun to form a relative clause

35. C

- Comfortably → comfortable

36. A

- This exercise → "this exercise" or "these exercises"

37. B

- Drinking → to drink

38. B

- Highest score → the highest score

39. C

- Went → had done

40. D

- it must be omitted to follow the pattern: something is too + adjective do do

PART ONE
題庫練習

PART TWO
模擬試卷

PART THREE
考生急症室

III. Sentence Completion

In this section, candidates are required to fill in the blanks with the best options given. The questions focus on grammatical use.

Example:

Complete the following sentence by choosing the best answer from the options given.

This market research company claims to predict in advance _____ by conducting exit polls of selected voters.

A. the results of an election will be

B. the results will be of an election

C. what results will be of an election

D. what the results of an election will be

E. what will the results of an election be

Answer : D

PART ONE
題庫練習
PART TWO
模擬試卷
PART THREE
考生急症室

Questions:

1. After centuries of obscurity, this philosopher's thesis is enjoying a surprising _____ .

   A. remission

   B. decimation

   C. longevity

   D. renaissance

   E. None of the above

2. There is a general _____ in the United States that our ethics are declining and that our moral standards are _____.

   A. feeling - normalizing

   B. idea - futile

   C. optimism - improving

   D. complaint - deteriorating

   E. None of the above

3. In the Middle Ages, the _____ of the great cathedrals did not enter into the architects' plans; almost invariably a cathedral was positioned haphazardly in _____ surroundings.

   A. situation - incongruous

   B. location - apt

C. ambience - salubrious

D. durability - convenient

E. None of the above

4. The quantum theory was initially regarded as absurd, unnatural and ____ with common sense.

    A. consanguineous

    B. discernible

    C. incompatible

    D. decipherable

    E. None of the above

5. Abraham Lincoln was famously known for his _____, which explains why he is often referred to as "Honest Abe."

    A. mendacity

    B. humility

    C. veracity

    D. piety

    E. None of the above

6. Archaeology is a poor profession; only ____ sums are available for excavating sites and even more ____

PART ONE
題庫練習

PART TWO
模擬試卷

PART THREE
考生急症室

amounts for preserving the excavations.

A. paltry - meager

B. miniscule - substantial

C. average - augmented

D. judicious - penurious

E. None of the above

7. The village headman was unlettered, but he was no fool, he could see through the _____ of the businessman's proposition and promptly _____ him down.

A. deception, forced

B. naivete, turned

C. potential, forced

D. sophistry, turned

E. None of the above

8. Moths are nocturnal pollinators, visiting scented flowers during the hours of darkness, whereas the butterflies are ____ , attracted to bright flowers in the daytime.

A. diurnal

B. quotidian

C. colorful

D. ephemeral

E. None of the above

9. War has been, throughout history, the chief _____ of social cohesion; and since science began, it has been the strongest _____ to technical progress.

   A. reason, encouragement

   B. origin, boost

   C. cause, provocation

   D. source, incentive

   E. None of the above

10. Scrooge, in the famous novel by Dickens, was a _____; he hated the rest of mankind.

    A. misanthrope

    B. hypochondriac

    C. philanthropist

    D. hedonist

    E. None of the above

11. In her crusade to make public servants accountable to voters, she_____ the nation's unscrupulous and

PART ONE
題庫練習

PART TWO
模擬試卷

PART THREE
考生急症室

self-indulgent politicians.

A. exposed

B. accepted

C. dramatized

D. promoted

E. None of the above

12. In a fit of _____ she threw out the valuable statue simply because it had belonged to her ex-husband.

A. pique

B. goodwill

C. contrition

D. pedantry

E. None of the above

13. In this biography we are given a glimpse of the young man _____ pursuing the path of the poet despite _____ and rejection slips.

A. doggedly - disappointment

B. sporadically - awards

C. tirelessly - encouragement

D. successfully - acclaim

E. None of the above

14. The assumption that chlorofluorocarbons would be _____ in the environment because they were chemically inert, was challenged by the demonstration of a potential threat to the ozone layer.

   A. deleterious

   B. innocuous

   C. persistent

   D. noxious

   E. None of the above

15. The professor became increasingly _____ in later years, flying into a rage whenever he was opposed.

   A. voluble

   B. subdued

   C. irascible

   D. contrite

   E. None of the above

16. People from all over the world are sent by their doctors to breathe the pure, ____ air in this mountain region.

   A. invigorating

   B. soporific

PART ONE
題庫練習
PART TWO
模擬試卷
PART THREE
考生急症室

C. debilitating

D. insalubrious

E. None of the above

17. Handicrafts constitute an important _____ of the decentralized sector of India's economy and _____ employment to over six million artisans.

A. factor, aims

B. extension, plants

C. segment, provides

D. period, projects

E. None of the above

18. Corruption is _____ in our society; the integrity of even senior officials is _____ .

A. rife - suspectful

B. growing - unquestioned

C. endangered - disputed

D. pervasive - intact

E. None of the above

19. Plastic bags are _____ symbols of consumer society; they are found wherever you travel.

    A. fleeting

    B. rare

    C. ephemeral

    D. ubiquitous

    E. None of the above

20. We live in a _____ age; everyone thinks that maximizing pleasure is the point of life.

    A. propitious

    B. sporadic

    C. corrupt

    D. hedonistic

    E. None of the above

21. Despite his _____ upbringing, Vladimir proved quite adept at navigating city life.

    A. urbane

    B. acrid

    C. bucolic

    D. cosmopolitan

    E. None of the above

PART ONE
題庫練習

PART TWO
模擬試卷

PART THREE
考生急症室

22. His presentation was so lengthy and _____ that it was difficult for us to find out the real _____ in it.

    A. verbose, content

    B. tedious, skill

    C. laborious, coverage

    D. simple, meaning

    E. None of the above

23. The _____ successfully repelled every _____ on the city.

    A. defenders, comment

    B. citizens, onslaught

    C. thieves, robbery

    D. judge, criticism

    E. None of the above

24. The success of the business venture ____ his expectations; he never thought that the firm would prosper.

    A. confirmed

    B. belied

    C. nullified

    D. fulfilled

    E. None of the above

25. Unwilling to admit that they had been in error, the researchers tried to _____ their case with more data obtained from dubious sources.

A. ascertain

B. buttress

C. refute

D. absolve

E. None of the above

26. Nutritionists declare that the mineral selenium, despite its toxic aspects, is_____to life, even though it is needed in extremely small quantities.

A. destructive

B. essential

C. insignificant

D. extraneous

E. None of the above

27. The formerly _____ waters of the lake have been polluted so that the fish are no longer visible from the surface.

A. muddy

B. tranquil

C. stagnant

D. pellucid

E. None of the above

28. Many so-called social playwrights are distinctly
_____ ; rather than allowing the members of the au-
dience to form their own opinions, these writers force
a viewpoint on the viewer.

A. conciliatory

B. prolific

C. iconoclastic

D. didactic

E. None of the above

29. A ____ child, she was soon bored in class; she already
knew more mathematics than her junior school teachers.

A. obdurate

B. querulous

C. precocious

D. recalcitrant

E. None of the above

30. The _____ terrorist was finally _____ by the police.

   A. famous, apprehended
   B. notorious, nabbed
   C. crafty, admonished
   D. renowned, caught
   E. None of the above

31. She hadn't eaten all day, and by the time she got home she was _____.

   A. blighted
   B. confutative
   C. ravenous
   D. ostentatious
   E. blissful

32. The movie offended many of the parents of its younger viewers by including unnecessary _____ in the dialogue.

   A. vulgarity
   B. verbosity
   C. vocalizations
   D. garishness
   E. tonality

PART ONE
題庫練習
PART TWO
模擬試卷
PART THREE
考生急症室

33. His neighbors found his _____ manner bossy and irritating, and they stopped inviting him to backyard barbeques.

    A. insentient

    B. magisterial

    C. reparatory

    D. restorative

    E. modest

34. Steven is always _____ about showing up for work because he feels that tardiness is a sign of irresponsibility.

    A. legible

    B. tolerable

    C. punctual

    D. literal

    E. belligerent

35. Candace would _____ her little sister into an argument by teasing her and calling her names.

    A. advocate

    B. provoke

    C. perforate

    D. lamente

    E. expunge

36. The dress Ariel wore _____ with small, glassy beads, creating a shimmering effect.

    A. titillated

    B. reiterated

    C. scintillated

    D. enthralled

    E. striated

37. Being able to afford this luxury car will _____ getting a better-paying job.

    A. maximize

    B. recombinant

    C. reiterate

    D. necessitate

    E. reciprocate

PART ONE
題庫練習

PART TWO
模擬試卷

PART THREE
考生急症室

38. Levina unknowingly _____ the thief by holding open the elevator doors and ensuring his escape.

    A. coerced

    B. proclaimed

    C. abetted

    D. sanctioned

    E. solicited

39. Shakespeare, a(n) _____ writer, entertained audiences by writing many tragic and comic plays.

    A. numeric

    B. obstinate

    C. dutiful

    D. prolific

    E. generic

40. I had the _____ experience of sitting next to an over-talkative passenger on my flight home from Brussels.

    A. satisfactory

    B. commendable

    C. galling

    D. acceptable

    E. acute

41. Prince Phillip had to choose: marry the woman he loved and _____ his right to the throne, or marry Lady Fiona and inherit the crown.

A. reprimand

B. upbraid

C. abdicate

D. winnow

E. extol

42. If you will not do your work of your own _____, I have no choice but to penalize you if it is not done on time.

A. predilection

B. coercion

C. excursion

D. volition

E. infusion

43. After sitting in the sink for several days, the dirty, food-encrusted dishes became _____.

A. malodorous

B. prevalent

C. imposing

D. perforated

E. emphatic

PART ONE
題庫練習
PART TWO
模擬試卷
PART THREE
考生急症室

44. Giulia soon discovered the source of the _____ smell in the room: a week-old tuna sandwich that one of the children had hidden in the closet.

    A. quaint

    B. fastidious

    C. clandestine

    D. laconic

    E. fetid

45. After making _____ remarks to the President, the reporter was not invited to return to the White House pressroom.

    A. hospitable

    B. itinerant

    C. enterprising

    D. chivalrous

    E. irreverent

46. With her _____ eyesight, Krystyna spotted a trio of deer on the hillside and she reduced the speed of her car.

A. inferior

B. keen

C. impressionable

D. ductile

E. conspiratorial

47. With a(n) _____ grin, the boy quickly slipped the candy into his pocket without his mother's knowledge.

A. jaundiced

B. nefarious

C. stereotypical

D. sentimental

E. impartial

48. Her _____ display of tears at work did not impress her new boss, who felt she should try to control her emotions.

A. maudlin

B. meritorious

C. precarious

D. plausible

E. schematic

49. Johan argued, "If you know about a crime but don't report it, you are _____ in that crime because you allowed it to happen."

A. acquitted

B. steadfast

C. tenuous

D. complicit

E. nullified

50. The authorities, fearing a _____ of their power, called for a military state in the hopes of restoring order.

A. subversion

B. premonition

C. predilection

D. infusion

E. inversion

51. The story's bitter antagonist felt such great _____ for all of the other characters that as a result, his life was very lonely and he died alone.

    A. insurgence

    B. malevolence

    C. reciprocation

    D. declamation

    E. preference

52. It is difficult to believe that charging 20% on an outstanding credit card balance isn't _____!

    A. bankruptcy

    B. usury

    C. novice

    D. kleptomania

    E. flagrancy

53. The _____ weather patterns of the tropical island meant tourists had to carry both umbrellas and sunglasses.

    A. impertinent

    B. supplicant

    C. preeminent

PART ONE
題庫練習

PART TWO
模擬試卷

PART THREE
考生急症室

D. illustrative

E. kaleidoscopic

54. Wedding ceremonies often include the exchange of _____ rings to symbolize the couple's promises to each other.

A. hirsute

B. acrimonious

C. plaintive

D. deciduous

E. votive

55. Kym was _____ in choosing her friends, so her parties were attended by vastly different and sometimes bizarre personalities.

A. indispensable

B. indiscriminate

C. commensurate

D. propulsive

E. indisputable

56. Phillip's _____ tone endeared him to his comical friends, but irritated his serious father.

  A. aloof

  B. jesting

  C. grave

  D. earnest

  E. conservative

57. Brian's pale Irish skin was _____ to burn if he spent too much time in the sun.

  A. prone

  B. urbane

  C. eminent

  D. erect

  E. daunted

58. A fan of historical fiction, Joline is now reading a novel about slavery in the _____ South.

  A. decorous

  B. rogue

  C. droll

  D. antebellum

  E. onerous

PART ONE
題庫練習
PART TWO
模擬試卷
PART THREE
考生急症室

59. Over the years the Wilsons slowly _____ upon the Jacksons' property, moving the stone markers that divided their lots farther and farther onto the Jacksons' land.

    A. encroached

    B. jettisoned

    C. conjoined

    D. repudiated

    E. teemed

60. Mary became _____ at typing because she practiced every day for six months.

    A. proficient

    B. reflective

    C. dormant

    D. redundant

    E. valiant

61. To find out what her husband bought for her birthday, Susan attempted to _____ his family members about his recent shopping excursions.

    A. prescribe

    B. probe

    C. alienate

    D. converge

    E. revere

62. Juan's friends found him in a _____ mood after he learned he would be homecoming king.

    A. jovial

    B. stealthy

    C. paltry

    D. gullible

    E. depleted

63. His suit of armor made the knight _____ to his enemy's attack, and he was able to escape safely to his castle.

    A. vulnerable

    B. churlish

    C. invulnerable

PART ONE
題庫練習

PART TWO
模擬試卷

PART THREE
考生急症室

D. static

E. imprudent

64. Choosing a small, fuel-efficient car is a _____ purchase for a recent college graduate.

A. corrupt

B. tedious

C. unhallowed

D. sardonic

E. judicious

65. Such a _____ violation of school policy should be punished by nothing less than expulsion.

A. copious

B. flagrant

C. raucous

D. nominal

E. morose

Answers:

1. D

The sentence tells us that the thesis has been in obscurity (forgotten or neglected) but now it is being revived. We can say it is undergoing a renaissance (revival).

2. D

The conjunction 'and' usually joins things of similar meaning or weight. This suggests that since ethics are declining, moral standards are also declining (deteriorating). Almost any word except 'optimism' would have fit the first blank.

3. A

The semicolon indicates that the second part of the sentence expands on the first part.

So, the second part tells us we are talking about the position, or situation of a cathedral. And since the first part tells us that architects did not pay attention to the situation, the cathedral was positioned randomly in odd (incongruous) surroundings.

4. C

A set of words linked with 'and' usually indicates things of similar weight. To go with 'absurd' and 'unnatural' we can choose incompatible with common sense.

5. C

Here, the keyword is "honest," and the missing word is used to explain why Lincoln was called honest. Thus, the missing word must also mean honesty, as only someone known for honesty would be called "honest.", As veracity means honesty or truthfulness, it's the suitable answer.

6. A

The part after the semicolon expands upon the first part of the sentence.

So, since the first part tells us that there is no money in archaeology, then there will only be small (paltry) amounts for excavating. Also 'even more' indicates that another similar word is required. Thus, meager also means small.

7. D

The sentence suggests that although the village headman was illiterate, he was wise. This information must have been provided in order to validate that he could see through the deception (sophistry) of the businessman's proposition and must have turned him down.

8. A

'Whereas' indicates opposite ideas.

PART **ONE**
題庫練習

PART **TWO**
模擬試卷

PART **THREE**
考生急症室

So we need a word in contrast to 'nocturnal' (active at night). The word is diurnal (active in daytime).

9. D

In the given space of time i.e. 'throughout history' there can be only source. Reason and cause or origin can be a particular event.

Boost, encouragement and incentive mean the same.

10. A

The part of the sentence after the semicolon defines the word needed for the blank.

So, since he 'hated mankind', the word we need is misanthrope (hater of the rest of mankind).

11. A

If the nation's politicians had been unscrupulous and self-indulgent, and a woman crusader wanted to make them accountable to the voters, she would not have dramatized, or accepted, or foreshadowed or promoted them, but would have exposed them.

12. A

She threw out a valuable statue just because it belonged to her ex-husband. Therefore, she must have been acting out of spite or ill will. Hence we choose pique which means 'resentment'.

13. A

The word 'despite' indicates something contrary. So, despite the rejection slips he pursues his path doggedly or tirelessly. But since the 'and' links similar things he is getting disappointment along with the rejection slips, not encouragement.

14. B

Turn the sentence around, and fit a word of your own in the blank. The demonstration of the potential threat challenged the assumption that the CFCs would be 'harmless' because they were inert. Innocuous is the best word because it means harmless.

15. C

The required word is signaled by the phrase ' flying into a rage'. Someone who is easily angered is described as 'irascible'.

16. A

Since the air is described as 'pure' we need a positive word. Also, since doctors recommend it, the air must be good for health.

Therefore, we choose invigorating which means energizing.

Soporific: Inducing sleep.

Debilitating: Causing a loss of energy or strength.

Insalubrious: Unhealthful.

17. C

The sentence says 'handicrafts constitute...'. The word 'constitute' suggests that the first blank will be filled with a word similar to part. From the given options, the only such word to fill in the blanks correctly would be 'segment'. The second blank would also be logically filled with the word 'provides'.

18. A

The semicolon suggests that the second part expands upon the first part. So, if corruption is rife (common), then we will doubt the integrity of the officials. Their integrity will be suspectful (doubtful).

19. D

Since the bags are found everywhere they must be ubiquitous.

20. D

The part after the semicolon explains what kind of age we are talking about.

So, since we are told that maximizing pleasure is the point, the word we need is hedonistic (pleasure seeking).

21. C

The keywords are found in the phrase "quite adept at navigating city life," which is used to describe Vladimir. The word 'despite' stresses that his ability to navigate city life is contrary to his upbringing. Hence, his upbringing must have been bucolic which is a word that suggests a rural lifestyle,

22. A

The word 'and' indicates that the first blank will be filled with a word that is similar to 'lengthy'. Hence, the first blank can be filled with 'verbose', 'tedious' and 'laborious'. The second blank can only be filled with the word 'content' to make complete meaning.

23. B

We can repel or push back only onslaughts or attacks.

24. B

He never thought the business would prosper (do well). Therefore, the success came as a surprise and contradicted ('belied') his expectations.

25. B

The researchers were unwilling to admit that they were wrong. Therefore they would try to support (buttress) their arguments.

PART ONE
題庫練習
PART TWO
模擬試卷
PART THREE
考生急症室

26. B

Something which is needed for life, even if it is in extremely small quantities, cannot be described as destructive to life, or insignificant to life, or extraneous to life. It can only be described as being essential to life.

27. D

The word 'formerly' indicates that once things were different. So, since now the waters are polluted so that fish cannot be seen, then formerly they must have been unpolluted and clear (pellucid).

28. D

The word for the blank is explained after the semicolon. If they force their viewpoint on the viewer, then they must be 'didactic'. Didactic means intending to preach or instruct.

29. C

The sentence suggests that she was way ahead of her class. This indicates the need for the word precocious, which means gifted or advanced for one's years.

30. B

The first blank describes a quality of the terrorist. This can be either filled with notorious or crafty as the other two options are positive words which cannot be used to describe a terrorist. The word 'admonished' means to scold firmly which does not suit the given context. Hence, the given blank has to be filled with the word 'nabbed'.

31. C

Ravenous means extremely hungry.

32. A

Vulgarity means offensive speech or conduct.

33. B

Magisterial means overbearing or offensively self-assured.

34. C

Punctual means arriving exactly on time.

35. B

To provoke is to incite anger or resentment; to call forth a feeling or action.

36. C

To scintillate means to emit or send forth sparks or little flashes of light, creating a shimmering effect; to sparkle.

37. D

To necessitate means to make necessary, especially as a result.

38. C

To abet means to assist, encourage, urge, or aid, usually an act of wrong-doing.

39. D

Prolific means abundantly creative.

40. C

Galling means irritating, annoying, or exasperating.

41. C

To abdicate means to formally relinquish or surrender power, office, or responsibility.

42. D

Volition means accord; an act or exercise of will.

43. A

Malodorous means having a foul-smelling odor.

44. E

Fetid means having a foul or offensive odor, putrid.

45. E

Irreverent means lacking respect or seriousness; not reverent.

46. B

Keen means being extremely sensitive or responsive; having strength of perception.

47. B

Nefarious means wicked, vicious, or evil.

48. A

Maudlin means excessively and weakly sentimental or tearfully emotional.

49. D

Complicit means participating in or associated with a questionable act or a crime.

50. A

Subversion means an overthrow, as from the foundation.

51. B

Malevolence means ill will or malice toward others; hate.

52. B

Usury is the lending of money at exorbitant interest rates.

PART ONE
題庫練習
PART TWO
模擬試卷
PART THREE
考生急症室

53. E

Kaleidoscopic means continually changing or quickly shifting.

54. E

Votive means dedicated by a vow.

55. B

Indiscriminate means not discriminating or choosing randomly; haphazard; without distinction.

56. B

Jesting means characterized by making jests; joking; playful.

57. A

Prone means a tendency or inclination to something.

58. D

Antebellum means belonging to the period before a war, especially the American Civil War.

59. A

To encroach means to gradually or stealthily take the rights or possessions of another; to advance beyond proper or formal limits; trespass.

60. A

Proficient means well versed in any business or branch of learning; adept.

61. B

To probe is to examine thoroughly; tentatively survey.

62. A

Jovial means showing hearty good cheer; marked with the spirit of jolly merriment.

63. C

Invulnerable means incapable of being damaged or wounded; unassailable or invincible.

64. E

Judicious means being wise or prudent; showing good judgment; sensible.

65. B

Flagrant means conspicuously and outrageously bad, offensive, or reprehensible.

PART **ONE**
題庫練習

PART **TWO**
模擬試卷

PART **THREE**
考生急症室

IV. Paragraph Improvement

In this section, two draft passages are cited. For each passage, questions are set to test candidates' skills in improving the draft. The focus of the questions is on writing skills, not power of understanding.

Examples :

The sentences below are parts of the early draft of two passages, some parts of which may have to be rewritten. Read the passages and choose the best answer to the question.

1. Which of the following versions of sentence (8) provides the best link between sentences (7) and (9), reproduced below?

   (7) It may be that a few of the products we have described are not available in some countries. (8) But it is possible to place an order via the Internet. (9) They will be dealt with promptly and efficiently.

   A. Furthermore, it is possible to order via the Internet.
   B. The Internet can be used in such circumstances.
   C. Orders can, however, be placed via the Internet.
   D. Sentence (8) as it is now. No change needed.

Answer : C

PART ONE
題庫練習

PART TWO
模擬試卷

PART THREE
考生急症室

2. Which of the following is the best revision of sentence (3), reproduced below?

(3) Mistakenly believing that smoking is a sign of maturity those in authority must act today to protect our citizens of tomorrow.

A. It is a mistake to believe that smoking is a sign of maturity. Those in authority must act today to protect our citizens of tomorrow.

B. It is a mistake to believe that smoking is a sign of maturity, those in authority must act today to protect our citizens of tomorrow.

C. Mistaken in their belief that smoking is a sign of maturity those in authority must act today to protect our citizens of tomorrow.

D. Those in authority should act today. Our citizens of today are mistakenly believing that smoking is a sign of maturity. They must be protected

Answer : A

Passage 1

(1) Have you ever went hang gliding? (2) Sailing smoothly across the sky, hang gliders are a sight to behold and capture humans' long-standing fascination with self propelled flight. (3) Whereas, it is also a dangerous pastime. (4) Personally, I know many people who are aware of the sport's relative danger but still enjoy it on a regular basis. (5) With proper technical training and certification, it is possible to avoid some of the most common hang gliding catamounts, stalling near the ground, aerobatic stunts gone wrong, and failure to use helmets or parachutes.

(6) In the United States, hang gliding is a relatively new sport and most hang gliders are wealthy and educated devotees. (7) Hang gliding actually originated in the 500s in China, with man-sized kites allowing people to sustain flight for short distances. (8) Continual advances in material technologies' result in constantly improving hang glider equipment – specialized nylon parachutes and lightweight aluminum alloy frames, for example. (9) Popular hang gliding destinations in the United States include Salt Lake City, Utah, Kitty Hawk, North Carolina, and Jackson Hole, Wyoming. (10) The most ardent affiliates say there's absolutely nothing like the joy of soaring miles above the ground for hours.

PART ONE
題庫練習
PART TWO
模擬試卷
PART THREE
考生急症室

1. **How should the tense error in Sentence 1 be corrected?**

   A. Will you ever have gone hang gliding?

   B. Have you ever went hang gliding? (no change)

   C. Have you ever gone hang gliding?

   D. Would you ever have went hang gliding?

   E. Would you ever have gone hang gliding?

Passage 2

(1) Have you ever went hang gliding? (2) Sailing smoothly across the sky, hang gliders are a sight to behold and capture humans' long-standing fascination with self propelled flight. (3) Whereas, it is also a dangerous pastime. (4) Personally, I know many people who are aware of the sport's relative danger but still enjoy it on a regular basis. (5) With proper technical training and certification, it is possible to avoid some of the most common hang gliding catamounts, stalling near the ground, aerobatic stunts gone wrong, and failure to use helmets or parachutes.

(6) In the United States, hang gliding is a relatively new sport and most hang gliders are wealthy and educated devotees. (7) Hang glid-

ing actually originated in the 500s in China, with man-sized kites allowing people to sustain flight for short distances. (8) Continual advances in material technologies' result in constantly improving hang glider equipment – specialized nylon parachutes and lightweight aluminum alloy frames, for example. (9) Popular hang gliding destinations in the United States include Salt Lake City, Utah, Kitty Hawk, North Carolina, and Jackson Hole, Wyoming. (10) The most ardent affiliates say there's absolutely nothing like the joy of soaring miles above the ground for hours.

2. What is the best way to revise the punctuation of Sentence 5?

A. With proper technical training and certification; it is possible to avoid some of the most common hang gliding catamounts, stalling near the ground, aerobatic stunts gone wrong, and failure to use helmets or parachutes.

B. With proper technical training and certification, it is possible to avoid some of the most common hang gliding catamounts: stalling near the ground, aerobatic stunts gone wrong, and failure to use helmets or parachutes.

C. With proper technical training and certification, it is possible to avoid some of the most common hang gliding catamounts. Stalling near the ground, aerobatic stunts gone wrong, and failure to use helmets or parachutes.

D. With proper technical training and certification, it is possible to avoid some of the most common hang gliding catamounts, stalling near the ground, aerobatic stunts gone wrong, and failure to use helmets or parachutes. (no change)

E. With proper technical training and certification, it is possible to avoid some of the most common hang gliding catamounts; stalling near the ground, aerobatic stunts gone wrong, and failure to use helmets or parachutes.

Passage 3

(1) Have you ever went hang gliding? (2) Sailing smoothly across the sky, hang gliders are a sight to behold and capture humans' long-standing fascination with self propelled flight. (3) Whereas, it is also a dangerous pastime. (4) Personally, I know many people who are aware of the sport's relative danger but still enjoy it on a regular basis. (5) With proper technical training and certification, it is possible to avoid some of the most common hang gliding catamounts, stalling near the ground, aerobatic stunts gone wrong, and failure to use

helmets or parachutes.

(6) In the United States, hang gliding is a relatively new sport and most hang gliders are wealthy and educated devotees. (7) Hang gliding actually originated in the 500s in China, with man-sized kites allowing people to sustain flight for short distances. (8) Continual advances in material technologies' result in constantly improving hang glider equipment – specialized nylon parachutes and lightweight aluminum alloy frames, for example. (9) Popular hang gliding destinations in the United States include Salt Lake City, Utah, Kitty Hawk, North Carolina, and Jackson Hole, Wyoming. (10) The most ardent affiliates say there's absolutely nothing like the joy of soaring miles above the ground for hours.

3. What is the best way to revise Sentence 2?

A. Sailing smoothly across the sky, hang-gliders are a sight to behold and capture humans' longstanding fascination with self propelled flight.

B. Sailing smoothly across the sky, hang gliders are a sight to behold and capture humans' longstanding fascination with self propelled flight.

C. Sailing smoothly across the sky, hang gliders are a sight to behold and capture humans' longstanding fascination with self propelled flight. (no

PART ONE
題庫練習

PART TWO
模擬試卷

PART THREE
考生急症室

change)

D. Sailing smoothly across the sky, hang gliders are a sight to behold and capture humans' longstanding fascination with self-propelled flight.

E. Sailing smoothly across the sky, hang-gliders are a sight to behold and capture humans' longstanding fascination with self-propelled flight.

Passage 4

(1) Have you ever went hang gliding? (2) Sailing smoothly across the sky, hang gliders are a sight to behold and capture humans' longstanding fascination with self propelled flight. (3) Whereas, it is also a dangerous pastime. (4) Personally, I know many people who are aware of the sport's relative danger but still enjoy it on a regular basis. (5) With proper technical training and certification, it is possible to avoid some of the most common hang gliding catamounts, stalling near the ground, aerobatic stunts gone wrong, and failure to use helmets or parachutes.

(6) In the United States, hang gliding is a relatively new sport and

most hang gliders are wealthy and educated devotees. (7) Hang gliding actually originated in the 500s in China, with man-sized kites allowing people to sustain flight for short distances. (8) Continual advances in material technologies' result in constantly improving hang glider equipment – specialized nylon parachutes and lightweight aluminum alloy frames, for example. (9) Popular hang gliding destinations in the United States include Salt Lake City, Utah, Kitty Hawk, North Carolina, and Jackson Hole, Wyoming. (10) The most ardent affiliates say there's absolutely nothing like the joy of soaring miles above the ground for hours.

4. How should Sentence 8 be rewritten?

A. Continual advances in material technologies' result in constantly improving hang glider equipment: specialized nylon parachutes and lightweight aluminum alloy frames, for example.

B. Continual advances in material technologies result in constantly improving hang glider equipment, specialized nylon parachutes and lightweight aluminum alloy frames, for example.

C. Continual advances in material technologies result in constantly improving hang glider equipment – specialized nylon parachutes and lightweight aluminum alloy frames, for example.

D. Continual advances in material technologies' re-

PART ONE
題庫練習
PART TWO
模擬試卷
PART THREE
考生急症室

sult in constantly improving hang glider equipment — specialized nylon parachutes and lightweight aluminum alloy frames, for example. (no change)

E. Continual advances in material technologies' result in constantly improving hang glider equipment; specialized nylon parachutes and lightweight aluminum alloy frames, for example.

Passage 5

(1) Although agritourism is a fairly recent phenomenon in the Western world; it is proving popular in many countries. (2) Agritourism is broadly defined as any activity or attraction that draws visitors to a farm, and it can include anything from corn mazes and apple picking to vineyard tours, workshops on animal husbandry, and workstay exchange programs. (3) The practice is particularly prevocalic in North America, Europe, and Australia. (4) With a huge variety of farms and activities available to agritourists. (5) For example: visitors can learn how to run a cattle drive in Wyoming, make cheese in France, harvest olives in Sicily, or pick kiwi fruit in New Zealand.

(6) Income generated from agritourism can help small family farms remain soluble as well as educate the public about where their food comes from. (7) Henceforth, most people agree that agritourism is benevolent for everyone involved. (8) While there are some who argue that it is a cheap ploy designed to make selfish tourists feel better about their vacations. (9) All in all, it will be interesting to see how agritourism continues to develop in the future.

5. How should Sentence 1 be rewritten?

A. Although agritourism is a fairly recent phenomenon in the Western world; it is proving popular in many countries. (no change)

B. Although agritourism is a fairly recent phenomenon in the Western world it is proving popular in many countries.

C. Although agritourism is a fairly recent phenomenon in the Western world, it is proving popular in many countries.

D. Although, agritourism is a fairly recent phenomenon in the Western world, it is proving popular in many countries.

E. Although, agritourism is a fairly recent phenomenon in the Western world; it is proving popular in many countries.

PART ONE
題庫練習

PART TWO
模擬試卷

PART THREE
考生急症室

Passage 6

(1) Although agritourism is a fairly recent phenomenon in the Western world; it is proving popular in many countries. (2) Agritourism is broadly defined as any activity or attraction that draws visitors to a farm, and it can include anything from corn mazes and apple picking to vineyard tours, workshops on animal husbandry, and work-stay exchange programs. (3) The practice is particularly prevocalic in North America, Europe, and Australia. (4) With a huge variety of farms and activities available to agritourists. (5) For example: visitors can learn how to run a cattle drive in Wyoming, make cheese in France, harvest olives in Sicily, or pick kiwi fruit in New Zealand.

(6) Income generated from agritourism can help small family farms remain soluble as well as educate the public about where their food comes from. (7) Henceforth, most people agree that agritourism is benevolent for everyone involved. (8) While there are some who argue that it is a cheap ploy designed to make selfish tourists feel better about their vacations. (9) All in all, it will be interesting to see how agritourism continues to develop in the future.

6. Which of the following is a sentence fragment?

A. Sentence 8

B. Sentence 7

C. Sentence 6

D. none of the other choices

E. Sentence 5

Passage 7

(1) Tattoos have even been found on ancient Icelandic, Egyptian, and South American mummies.(2) On their various voyages to the South Pacific, Captain Cook and other European explorers brought back accounts of colorfully inked natives, and their sailors soon began to adopt the practice. (3) Early medieval Northern European tribes such as the Picts and Visigoths were often heavily tattooed; particularly the warriors. (4) Over the centuries, various cultures have described various meanings to tattoos, with indelible ink signifying everything from royalty to gang membership to magical powers.

(5) Interestingly modern-day tattooing's popularity evolved out of

PART ONE
題庫練習

PART TWO
模擬試卷

PART THREE
考生急症室

its adoption by wealthy British nobility, and by American celebrities, musicians, and actors. (6) While tattoos do have a dark past, most notably at Auschwitz, where Nazi's identified prisoners by tattooing numbers on their arms. (7) Tattoos can be signs of joy, belief, or even healing, such as the colorful tattoos that breast cancer survivors use to cover mastectomy scars. (8) Today, there are more options than ever before, tattoo artists often have fine arts training, lengthy apprenticeships, and can offer specialty inks that glow in the dark or are easier to remove. (9) The internet is full of design ideas, tattoo parlor rankings, and even horror stories about bad tattoo experiences. (10) Now more than ever, information about ink abounds. (11) Choose wisely!

7. Which of the following sentences needs an apostrophe deleted from it?

A. Sentence 3

B. Sentence 7

C. Sentence 6

D. Sentence 5

E. Sentence 4

Passage 8

(1) Some of the best-known badlands occur in North America; Badlands National Park in South Dakota, Theodore Roosevelt National Park in North Dakota, Dinosaur National Monument in Colorado and Utah, and the Big Muddy Badlands in Saskatchewan. (2) Have you ever been to the badlands?

(3) Despite their name, badlands are often fascinating topographical regions. (4) With beautiful vistas as well. (5) Elsewhere badlands can be found in Italy, New Zealand, Spain, and Argentina. (6) Badlands are distinguished by their stark, dry terrain, their sharply eroded landscapes, their lack of vegetation, and their colorful, clay-rich rocks. (7) They often include geological features such as canyons, gullies, mesas, buttes, and hoodoos. (8) Nevertheless, visiting badlands can be an uncommon but rewarding experience.

8. What punctuation mark should be added to Sentence 6 to provide clarity?

A. none of the other answers

B. Semicolon (;)

PART ONE
題庫練習

PART TWO
模擬試卷

PART THREE
考生急症室

C. Comma (,)

D. Colon (:)

E. Hyphen (–)

Passage 9

(1) While many people feel strongly about the genetic modification of foods. (2) Most do not understand the full nature of genetically modified (GM) plants and animals. (3) For example, did you even know that GM was the abbreviation for genetically modified? (4) Genetic modification is defined as the artificial manipulation of a specie's DNA usually for the purpose of improving crop yield, resistance to disease, or nutritive value. (5) And did you know it has been going on for millennia, because ancient farmers were selectively breeding foods to provide better flavor, produce smaller seeds, or withstand drought and flooding?

(6) Today, however, having more sophisticated tools allow scientists to transfer genes from one organism with another, this purposely provokes better pest resistance and other desirable characteristics. (7) Opponents of genetic modification argue that the DNA modifications

are unstable, undesirable, and unhealthy for the environment. (8) Additionally, some are also arguing that consuming GM foods is unsafe for humans, despite much scientific evidence to the contrary.

9. How should Sentence 4 be rewritten?

A. Genetic modification is defined as the artificial manipulation of a species' DNA, usually for the purpose of: improving crop yield, resistance to disease, or nutritive value.

B. Genetic modification is defined as the artificial manipulation of a species' DNA, usually for the purpose, of improving crop yield, resistance to disease, or nutritive value.

C. Genetic modification is defined as the artificial manipulation of a species' DNA; usually for the purpose of improving crop yield, resistance to disease, or nutritive value.

D. Genetic modification is defined as the artificial manipulation of a species' DNA usually for the purpose of improving crop yield, resistance to disease, or nutritive value. (no change)

E. Genetic modification is defined as the artificial manipulation of a species' DNA, usually for the purpose of improving crop yield, resistance to disease, or nutritive value.

PART ONE
題庫練習
PART TWO
模擬試卷
PART THREE
考生急症室

Passage 10

(1) You may not know Gerard Manley Hopkins was a famous English poet. (2) Hopkins led a complicated life as a Jesuit priest, converting to Roman Catholicism in 1866. (3) Born in 1844 the poet was excellent at sketching from an early age and attended the University of Oxford from 1863 to 1867, where he met poets Christina Rossetti, Robert Bridges, and others. (4) According to his personal diaries, Hopkins frequently struggled to repress homoerotic urges, adopting an ascetic lifestyle, many believing that this contributed to his writing. (5) His work itself is characterized by an escarpment of conventional poetic meter, the use of sprung rhythm, frequent vivid imagery, and a careful and creative use of language.(6) Sprung rhythm is a particular poetic rhythm that is intended to mimic natural speech and is distinguished by its irregular patterns although it is distinct from free verse.

(7) Hopkins died when he was only in his forties, but his contributions to poetry – particularly his experimentation and his use of sprung rhythm – continue to obscure today.

10. How should Sentence 6 be rewritten?

A. Sprung rhythm, a particular poetic rhythm that is intended to mimic natural speech, is distinguished by its irregular patterns but is distinct from free verse.

B. Sprung rhythm is a particular poetic rhythm that is intended to mimic natural speech and is distinguished by its irregular patterns although it is distinct from free verse. (no change)

C. Sprung rhythm is a particular poetic rhythm, which, intended to mimic natural speech, is distinguished by its irregular patterns, because it is distinct from free verse.

D. Sprung rhythm is a particular poetic rhythm, it is intended to mimic natural speech, and is distinguished by its irregular patterns, while it is distinct from free verse.

E. Sprung rhythm: a particular poetic rhythm that is intended to mimic natural speech, is distinguished by its irregular patterns although it is distinct from free verse.

PART ONE
題庫練習

PART TWO
模擬試卷

PART THREE
考生急症室

Passage 11

(1) Punk rock developed in the mid-1970s. (2) It was a musical movement that arose out of antiauthoritarian garage bands.(3) It was characterized by fast-paced songs, sedimentary lyrics, and a raw loud sound. (4) And often its lyrics were also political. (5) Some of the most famous punk rock bands came from England and the United States and including the Clash, the Sex Pistols, and the Ramones.

(6) Punk bands tending to convince a liberal, anti-establishment, sensibility,and they were proponents of individualism, freedom, and nonconformity. (7) (Later in the 1990s "riot grrrl" bands like Bikini Kill and Sleater-Kinney used their punk music to draw attention on feminist concerns.) (8) Now you can find, punk bands in cities all around the world. (9) By the 1980s, the public was beginning to accept punk music, slowly becoming mainstream.

11. How should Sentence 8 be rewritten?

A. Now you can find: punk bands in cities all around the world.

B. Now you can find, punk bands in cities all around the world. (no change)

C. Now, you can find, punk bands in cities all around the world.

D. Now, you can find: punk bands in cities all around the world.

E. Now you can find punk bands in cities all around the world.

Passage 12

(1) If you've watched any environmental news reports in the last few years, it's likely you've stumbled among the idea of fracking. (2) The word is short for "hydraulic fracturing" and involves injecting liquid into rock to create fractures and fractals, there allowing natural gas to be extracted more querulously. (3) Proponents say the method facilitates oil drilling and allows countries, like the United States, to cut back on their foreign oil dependence. (4) Amateurs say that fracking, causes significant and sometimes irresponsible environmental damage.

(5) Fracking can require sonorous quantities of water, and leach dangerous carcinogenic chemicals into the groundwater. (6) Some

people have even inquired minor earthquakes to fracking: as the process thought to spurn tremors.(7) Perhaps most worrisome, fracking allows governments to continue depending on fossil fuel rather than exploring renewable energy. (8) These sources could include wind turbines, solar panels, even hot springs and waterwheels.

12. How should Sentence 5 be rewritten?

A. Fracking can require sonorous quantities of water; and leach dangerous carcinogenic chemicals into the groundwater.

B. Fracking can require sonorous quantities of water; leach dangerous carcinogenic chemicals into the groundwater.

C. Fracking can require sonorous quantities of water, leach dangerous carcinogenic chemicals into the groundwater.

D. Fracking can require sonorous quantities of water, leaching dangerous carcinogenic chemicals into the groundwater.

E. Fracking can require sonorous quantities of water, and leach dangerous carcinogenic chemicals into the groundwater. (no change)

Passage 13

(1) Thoreau went to them when he wished to live deliberately, Whitman wrote about them in his book Leaves of Grass. (2) They clean the air, and harbor wildlife. (3) There are nearly 200 million, protected acres of them in the United States. (4) Forests are a national treasure.

(5) Since the late 1800s, Congress and Presidents have decorated various areas of the country as national forests, grasslands, and preserves. (6) Some well known National Parks include the Grand Canyon, Yosemite, Yellowstone, and the Great Smoky Mountains. (7) Protecting forests helps safeguard native plant and animal species, it honors landscapes complex histories, and preserving the natural beauty for future generations.

13. How should Sentence 7 be rewritten?

A. Protecting forests helps safeguard native plant and animal species, honor landscapes complex histories, and preserves the natural beauty for future generations.

B. Protecting forests, which helps safeguard na-

PART ONE
題庫練習

PART TWO
模擬試卷

PART THREE
考生急症室

tive plant and animal species, honor landscapes complex histories, preserves the natural beauty for future generations.

C. Protecting forests helps safeguarding native plant and animal species, honoring landscapes complex histories, and preserving the natural beauty for future generations.

D. Protecting forests helps safeguard native plant and animal species, honor landscapes complex histories, and preserve the natural beauty for future generations.

E. Protecting forests helps safeguard native plant and animal species, it honors landscapes complex histories, and preserving the natural beauty for future generations. (no change)

Passage 14

(1) Some of the best-known badlands occur in North America; Badlands National Park in South Dakota, Theodore Roosevelt National Park in North Dakota, Dinosaur National Monument in Colorado and Utah, and the Big Muddy Badlands in Saskatchewan. (2) Have you ever been to the badlands? (3) Despite their name, badlands are often

fascinating topographical regions. (4) With beautiful vistas as well.

(5) Elsewhere badlands can be found in Italy, New Zealand, Spain, and Argentina.(6) Badlands are distinguished by their stark, dry terrain, their sharply eroded landscapes, their lack of vegetation, and their colorful, clay-rich rocks. (7) They often include geological features such as canyons, gullies, mesas, buttes, and hoodoos. (8) Nevertheless, visiting badlands can be an uncommon but rewarding experience.

14. What is the best way to combine Sentences 3 and 4?

A. Despite their name, badlands are often fascinating topographical regions, which are including beautiful vistas as well.

B. Despite their name, badlands are often fascinating topographical regions, as well having beautiful vistas.

C. Despite their name, badlands are often fascinating topographical regions with beautiful vistas.

D. Despite their name, badlands are often fascinating topographical regions. With beautiful vistas as well. (no combination)

E. Despite their name, badlands are often fascinating topographical regions, that including beautiful vistas as well.

PART ONE
題庫練習

PART TWO
模擬試卷

PART THREE
考生急症室

Passage 15

(1) Punk rock developed in the mid-1970s. (2) It was a musical movement that arose out of antiauthoritarian garage bands.(3) It was characterized by fast-paced songs, sedimentary lyrics, and a raw loud sound. (4) And often its lyrics were also political. (5) Some of the most famous punk rock bands came from England and the United States and including the Clash, the Sex Pistols, and the Ramones.

(6) Punk bands tending to convince a liberal, anti-establishment, sensibility,and they were proponents of individualism, freedom, and nonconformity.(7) (Later in the 1990s "riot grrrl" bands like Bikini Kill and Sleater-Kinney used their punk music to draw attention on feminist concerns.) (8) Now you can find, punk bands in cities all around the world. (9) By the 1980s, the public was beginning to accept punk music, slowly becoming mainstream.

15. Where should Sentence 8 be moved?

A. After Sentence 2

B. It should not be moved

C. After Sentence 9

D. After Sentence 5

E. After Sentence 6

Answers

1. C

While several of these options are grammatically correct, the version that best preserves the original meaning of the sentence is "Have you ever gone hang-gliding?"

2. B

Here, a colon is the optimal punctuation to introduce a list after an independent clause.

3. D

Here, "hang gliders" should not be hyphenated (it's simply a two-word noun), "longstanding" is one word, not two, and "self propelled" should be hyphenated, as with most words that have a "self-" prefix.

4. C

Here, we have an apostrophe error. "Technologies" is not a possessive word in this case, so it should not have an apostrophe. Either a colon or an em dash is fine to separate "equipment" and "specialized."

5. C

In the original sentence, a semicolon is incorrectly separating a dependent clause from an independent clause. The correct punctuation is a comma. Generally speaking, no comma is necessary after the word "although."

6. A

Sentence 8 is a dependent clause lacking an independent clause, which makes it a sentence fragment. Here is one way to rewrite it correctly: "However, there are some who argue that it is a cheap ploy designed to make selfish tourists feel better about their vacations."

7. C

The plural of Nazi is Nazis, not "Nazi's". (Generally, apostrophes are not used to change a singular noun into a plural one.)

8. B

Sentence 6 presents complex items in a list. Because these items themselves contain commas, adding a semicolon as a sort of "super comma" keeps the items in the list distinct from each other. The correction would look like this: "Badlands are distinguished by their stark, dry terrain; their sharply eroded landscapes; their lack of vegetation; and their colorful, clay-rich rocks."

9. E

Here, we have an independent clause

PART ONE
題庫練習
PART TWO
模擬試卷
PART THREE
考生急症室

("Genetic modification is defined as the artificial manipulation of a species' DNA") and a dependent clause ("usually for the purpose of improving crop yield, resistance to disease, or nutritive value") that must be separated with a comma.

10. A

The most concise sentence makes several small grammatical changes.

11. E

No comma or colon is necessary in the sentence; in fact, adding punctuation would impede the reader's progress through this single, unified thought and break up the sentence unnecessarily.

12. D

A comma should only be used before the coordinating conjunction "and" if it separates two independent clauses. Changing the verb tense in the second (dependent) clause and deleting the conjunction allows us to keep the comma. Note also that semicolons cannot be used to connect a dependent clause to a main clause and that coordinating conjunctions cannot follow semicolons.

13. D

Remember that lists should always

exemplify proper parallelism, and be very careful to match all the verb tenses. In the example sentence each item takes a different form, one extraneously including "it" as a subject, the other using an unnecessary and incorrect gerund.

14. C

The most concise and grammatically correct option is this: "Despite their name, badlands are often fascinating topographical regions with beautiful vistas."

15. C

Sentence 8 serves as a good conclusion for the passage, since it is the most contemporary sentence in a passage arranged chronologically. It does not make sense to include a summing up of where the punk rock movement is "now," and then jump back in time to the 1980s.

模擬試卷一

Mock Paper 1

Time Limit: 45 minutes

PART ONE
題庫練習

PART TWO
模擬試卷

PART THREE
考生急症室

I. Comprehension (10 questions)

Read the following passage and answer Questions 1-10. For each question, select the best answer from the given alternatives.

The University Grants Commission's directive to college and university lecturers to spend a minimum of 22 hours a week in direct teaching is the product of budgetary cutbacks rather than pedagogic wisdom. It may seem odd, at first blush, that teachers should protest about teaching a mere 22 hours. However, if one considers the amount of time academics require to prepare lectures of good quality as well as the time they need to spend doing research, it is clear that most conscientious teachers work more than 40 hours a week. In university systems around the world, lecturers rarely spend more than 12 to 15 hours in direct teaching activities a week. The average college lecturer in India does not have any office space. If computers are available, internet connectivity is unlikely. Libraries are poorly stocked. Now the UGC says universities must implement a complete freezeon all permanent recruitment, abolish all posts which have been vacant for more than a year, and cut staff strength by 10 per cent. And it is in order to ensure that these cutbacks do not affect the quantum of teaching that existing lecturers are being asked to work longer. Obviously, the quality of teaching and academic work in general will decline. While it is true that some college teachers do not take their classes regularly, the UGC and the Institutions concerned must find a proper way to hold them accountable. An absentee teacher will continue to play truant even if the number of hours he is required to teach goes up.

All of us are well aware of the unsound state that the Indian higher education system is in today. Thanks to years of sustainedfinancial neglect, most Indian universities and colleges do no research worth the name. Even as the number of students entering colleges has increased dramatically, public investment in higher education has actually declined in relative terms. Between 1985 and 1997, when public expenditure on higher education as a percentage of outlays on all levels of education grew by more than 60 per cent in Malaysia and 20 per cent in Thailand. India showed a decline of more than 10 per cent. Throughout the world, the number of teachers in higher education per million populationgrew by more than 10 per cent-in the same period; in India it fell by one per cent. Instead of transferring the burden of government apathy on to the' backs of the teachers, the UGC should insist that the needs of the country's university system, are adequately catered for.

PART ONE
題庫練習

PART TWO
模擬試卷

PART THREE
考生急症室

1. Why does the UGC want to increase the direct teaching hours of university teachers?

   A. UGC feels that the duration of contact between teacher and the taught should be more.

   B. UGC wants teachers to spend more time in their departments.

   C. UGC wants teachers to devote some time to improve university administration.

   D. UGC does not have money to appoint additional teachers.

   E. None of these

2. Which of the following is the reason for the sorry state of affairs ofthe Indian Universities as mentioned in the passage?

   A. The poor quality of teachers

   B. Involvement of teachers in extra-curricular activities

   C. Politics within and outside the departments

   D. Heavy burden of teaching hours on the teachers

   E. Not getting enough financial Assistance

3. Which of the following statements/is/are TRUE in the context of the passage?

    (A) Most colleges do not carry out research worth the name.

    (B) UGC wants lecturers to spend minimum 22 hours a week in direct teaching.

    (C) Indian higher education system is in unsound state.

    A. Only (A)and (C)
    B. All (A) (B) and (C)
    C. Only (C) (d)Only (B)
    D. Only (B)
    E. Only (B)and ©

4. Besides direct teaching University teachers spend considerable time in/on .

    A. Administrative activities such as admissions
    B. Supervising examinations and correction of answer papers
    C. Carrying out research in the area of their interest
    D. Maintaining research equipment and libraries
    E. Developing liaison with the user organisations

PART ONE
題庫練習

PART **TWO**
模擬試卷

PART THREE
考生急症室

5. Which of the following statements is NOT TRUE in the context of the passage?

    A. UGC wants teachers to spend minimum 40 hours in a week in teaching

    B. Some college teachers do not engage their classes regularly

    C. The average college teacher in India does not have any office space

    D. UGC wants universities to abolish all posts which have been vacant for more than a year

    E. All are true

6. Between 1985- 1997, the number of teachers in higher education per million population, in India

    A. Increased by 60%

    B. Increased by 20%

    C. Decreased by 22%

    D. Decreased by 10%

    E. Decreased by 1%

7. Which of the following statements is NOT TRUE in the context of the passage?

    A. Indian universities are financially neglected.

    B. All over the world, the university lecturers hardly spend more than 12 to 15 hours a week in direct teaching.

    C. Indian Universities are asked to reduce staff strength by 10%.

    D. Public investment in higher education has increased in India.

    E. Malaysia spends more money on education than Thailand.

8. What is the UGC directive to the universities?

    A. Improve the quality of teaching.

    B. Spend time on research activities.

    C. Do not appoint any permanent teacher.

    D. Provide computer and internet facilities.

    E. Do not spend money on counseling services to the students.

PART ONE
題庫練習

PART TWO
模擬試卷

PART THREE
考生急症室

9. Identify the correct synonym for the word Flush from the passage.

   A. Freeze

   B. Sustain

   C. Blush

   D. All of these

   E. None of these

10. Suggest a suitable replacement for the word in the line which is highlighted.... 'And it is in order to ensure that these cutbacks do not affect the quantum of teaching that existing lecturers are being asked to work longer.'

   A. Totality

   B. Amount

   C. Persona

   D. Physics

   E. None of these

II. Error Identification (10 questions)

The sentences below may contain a language error. Identify the part that contains the error of choose "(E) No error" if the entire sentence is correct.

11. <u>A secretary</u> told me <u>important file</u> <u>had been left</u> in the lunch room just <u>the other day</u>.

    A. A secretary
    B. important file
    C. had been left
    D. the other day
    E. No Error

12. <u>Having punished</u> twice <u>this</u> week, Kate feels <u>ashamed of</u> her bad <u>behavior</u>.

    A. Having punished
    B. this
    C. ashamed of
    D. behaviour
    E. No Error

PART ONE
題庫練習
PART TWO
模擬試卷
PART THREE
考生急症室

13. The WWF workers <u>have</u> devoted <u>all</u> their lives to <u>protect</u> all <u>animals</u>.

    A. have

    B. all

    C. protect

    D. animals

    E. No Error

14. The <u>more careful</u> you drive, <u>the fewer accidents</u> you <u>will have</u>.

    A. more careful

    B. the fewer

    C. accidents

    D. will have

    E. No Error

15. Nobody in this class <u>know anything about</u> the <u>changes</u> in the <u>exam formats</u>.

    A. know

    B. anything about

    C. changes

    D. exam formats

    E. No Error

16. Peter <u>couldn't stay</u> <u>on</u> the <u>horse's back</u> and <u>neither Bob could</u>.

    A. couldn't stay

    B. on

    C. horse's back

    D. neither Bob could

    E. No Error

17. <u>The more</u> I live <u>with</u> him, <u>the most</u> I <u>love him</u>.

    A. The more

    B. with

    C. the most

    D. love him

    E. No Error

18. <u>Get</u> a builder <u>put</u> in a lift and then you <u>won't</u> have to climb <u>up</u> all these stairs.

    A. Get

    B. put

    C. won't

    D. up

    E. No Error

PART ONE
題庫練習
PART **TWO**
模擬試卷
PART THREE
考生急症室

19. <u>I</u> <u>feel</u> <u>safely</u> to <u>tell him my secrets</u>.

    A. I

    B. feel

    C. safely

    D. tell him my secrets

    E. No Error

20. Nonla, <u>that</u> is one <u>of</u> the <u>typical</u> symbols of the Vietnamese culture, has <u>a</u> conical form.

    A. a

    B. of

    C. that

    D. typical

    E. No Error

III. Sentence Completion (10 questions)

Complete the following sentences by choosing the best answer from the options given.

21. As _____ beings we live each day conscious of our shortcomings and victories.

    A. sensational

    B. sentient

    C. sentimental

    D. static

    E. senile

22. The curious crowd gathered to watch the irate customer _____ about the poor service he received in the restaurant.

    A. antiquate

    B. trivialize

    C. rant

    D. placate

    E. fetter

PART ONE
題庫練習

PART **TWO**
模擬試卷

PART THREE
考生急症室

23. The man's _____ driving resulted in a four-car pile-up on the freeway.

    A. burdensome

    B. charismatic

    C. exceptional

    D. boastful

    E. negligent

24. Ron didn't know the rules of rugby, but he could tell by the crowd's reaction that it was a critical _____ in the game.

    A. acclamation

    B. conviction

    C. juncture

    D. enigma

    E. revelation

25. My ancestor who lost his life in the Revolutionary War was a _____ for American independence.

    A. knave

    B. reactionary

    C. compatriot

    D. nonconformist

    E. martyr

26. The _____ sound of the radiator as it released steam became an increasingly annoying distraction.

   A. sibilant

   B. scintillating

   C. diverting

   D. sinuous

   E. scurrilous

27. It is helpful for salesmen to develop a good _____ with their customers in order to gain their trust.

   A. platitude

   B. rapport

   C. ire

   D. tribute

   E. disinclination

28. In such a small office setting, the office manager found he had _____ responsibilities that required knowledge in a variety of different topics.

   A. heedless

   B. complementary

   C. mutual

PART ONE
題庫練習

PART TWO
模擬試卷

PART THREE
考生急症室

D. manifold

E. correlative

29. David's _____ entrance on stage disrupted the scene and caused the actors to flub their lines.

A. untimely

B. precise

C. lithe

D. fortuitous

E. tensile

30. The settlers found an ideal location with plenty of _____ land for farming and a mountain stream for fresh water and irrigation.

A. candid

B. provincial

C. arable

D. timid

E. quaint

IV. Paragraph Improvement

(10 questions)

The sentences below are parts of early drafts of one or more passages, some parts of which may have to be rewritten. Read the draft and choose the best answer to the question.

Passage A

(1) The early history of astronomy was full of misunderstandings. (2) Some of them were funny, it's likethe controversy of the "canali" on Mars. (3) In the late 1800's an Italian astronomer named Giovanni Schiaparellistudied Mars. (4) He had a high-powered telescope that hused to look at Mars. (5) Schiaparelli thought he saw channels crisscrossing the planet's surface. (6) He was intrigued: perhaps these channels were evidence that Marhad great flowing rivers like the Earth. (7) Schiaparelli made charts of the surface of Mars and labeled it with theItalian word "canali."

(8) Unfortunately, "canali" can be translated into English as either "channels" or "canals." (9) Channels andcanals are two different things because channels are formnaturally by water, while canals are constructed by people. (10) Some people translated "canali" as "ca-

PART ONE
題庫練習

PART **TWO**
模擬試卷

PART THREE
考生急症室

nals," word began to spread that the lines Schiaparelli saw through histelescope were actually canals that had been built by intelligent beings. (11) One of them was an amateur astronomer named Percival Lowell. (12) He wrote a serieof best-selling books. (13) In these books Lowell publicized the notion that these "canals" were built by Martian farmers who understood irrigation.

(14) In 1965 a U.S. spacecraft flying close to the surface of Mars sent back conclusive pictures. (15) There are no prominent channels anywhere on the planet. (16) Lowell and Schiaparelli saw what they wanted to see. (17) Lowell was wrong, of course, but so was Schiaparelli.

31. Which is the best version of sentence 2 reproduced below?

    A. as it is now

    B. Some of them were funny; it's like the controversy of the "canali" on Mars.

    C. Some of them were funny, like the controversy of the "canali" on Mars.

    D. Some of them were funny, even the controversy of the "canali" on Mars.

    E. Some of them were as funny as the controversy of the "canali" on Mars.

32. Which is the best way to combine sentences 3 and 4 reproduced below?

   A. In the late 1800's an Italian astronomer named Giovanni Schiaparelli studied Mars by a high-powered telescope.

   B. In the late 1800's an Italian astronomer named Giovanni Schiaparelli studied Mars with a high-powered telescope that he used to look at Mars.

   C. In the late 1800's an Italian astronomer named Giovanni Schiaparelli studied Mars, he had a high-powered telescope that he used.

   D. In the late 1800's an Italian astronomer named Giovanni Schiaparelli used a high-powered telescope to study Mars.

   E. In the late 1800's an Italian astronomer named Giovanni Schiaparelli, using a high-powered telescope to look, studied Mars.

33. Which word would be best to insert at the beginning of sentence 10 reproduced below?

   A. Whereas

   B. However

   C. If

   D. Although

   E. Because

PART ONE
題庫練習

PART TWO
模擬試卷

PART THREE
考生急症室

34. What is the best version of sentence 11 reproduced below?

A. As it is now

B. One of the most intelligent was an amateur astronomer named Percival Lowell.

C. This idea was popularized by an amateur astronomer named Percival Lowell.

D. It was spread even more by someone else, an amateur astronomer named Percival Lowell.

E. The person who solved the problem was an amateur astronomer named Percival Lowell.

35. What is the best way to combine sentences 12 and 13 reproduced below?

A. In a series of bestselling books, Lowell publicized the notion that these "canals" were built by Martian farmers who understood irrigation.

B. He wrote a series of books that was a bestseller and publicized the notion that these "canals" were built by Martian farmers who understood irrigation.

C. His books that were bestsellers publicized the notion that these "canals" were built by Martian farmers who understood irrigation.

D. He wrote a series of bestselling books; Lowell publicized the notion that these "canals" were built by Martian farmers who understood irrigation.

E. In these books, which were bestsellers, Lowell publicized the notion that these "canals" were built by Martian farmers who understood irrigation.

Passage B

(1) Not many children leave elementary school and they have not heard of Pocahontas' heroic rescue of John Smith from her own people, the Powhatans. (2) Generations of Americans have learned the story of a courageous Indian princess who threw herself between the Virginia colonist and the clubs raised to end his life. (3) The captive himself reported the incident. (4) According to that report, Pocahontas held his head in her arms and laid her own upon his to save him from death.

(5) But can Smith's account be trusted? (6) Probably it cannot, say several historians interested in dispelling myths about Pocahontas.

PART ONE
題庫練習

PART TWO
模擬試卷

PART THREE
考生急症室

(7) According to these experts, in his eagerness to find patrons for future expeditions, Smith changed the facts in order to enhance his image. (8) Portraying himself as the object of a royal princess' devotion may have merely been a good public relations ploy. (9) Research into Powhatan culture suggests that what Smith described as an execution might have been merely a ritual display of strength. (10) Smith may have been a character in a drama in which even Pocahontas was playing a role.

(11) As ambassador from the Powhatans to the Jamestown settlers, Pocahontas headed off confrontations between mutually suspicious parties. (12) Later, after her marriage to colonist John Rolfe, Pocahontas traveled to England, where her diplomacy played a large part in gaining support for the Virginia Company.

36. What is the best way to deal with sentence 1 reproduced below?

   A. Leave it as it is.
   B. Switch its position with that of sentence 2.
   C. Change "leave" to "have left".
   D. Change "and they have not heard" to "without having heard".
   E. Remove the comma and insert "known as the".

37. In context, which of the following is the best way to revise the wording in order to combine sentences 3 and 4 ?

   A. The captive himself reported the incident, according to which Pocahontas held his head in her arms and laid her own upon his to save him from death.

   B. Since then, the captive reported the incident, which said that Pocahontas held his head in her arms and laid her own upon his to save him from death.

   C. Consequently, the captive himself reports that Pocahontas held his head in her arms and laid her own upon his to save him from death.

   D. It seems that in the captive's report of the incident he says that Pocahontas held his head in her arms and laid her own upon his to save him from death.

   E. According to the captive's own report of the incident, Pocahontas held his head in her arms and laid her own upon his to save him from death.

38. Which of the following phrases is the best to insert at the beginning of sentence 10 to link it to sentence 9?

   A. Far from being in mortal danger,

PART ONE
題庫練習

PART **TWO**
模擬試卷

PART THREE
考生急症室

B. If what he says is credible,

C. What grade school history never told you is this:

D. They were just performing a ritual, and

E. But quite to the contrary,

39. Which of the following best describes the relationship between sentences 9 and 10?

A. Sentence 10 concludes that the theory mentioned in sentence 9 is wrong.

B. Sentence 10 adds to information reported in sentence 9.

C. Sentence 10 provides an example to illustrate an idea presented in sentence 9.

D. Sentence 10 poses an argument that contradicts the point made in sentence 9.

E. Sentence 10 introduces a new source that confirms the claims made in sentence 9.

40. Which of the following would be the best sentence to insert before sentence 11 to introduce the third paragraph?

A. It is crucial to consider the political successes as well as the shortcomings of Pocahontas.

B. The Pocahontas of legend is the most interesting, but the historical Pocahontas is more believable.

C. If legend has overemphasized the bravery of Pocahontas, it has underplayed her political talents.

D. To really know Pocahontas, we must get beyond myth and legend to the real facts about her private life.

E. Perhaps we will never really know the real Pocahontas.

END OF PAPER

PART ONE
題庫練習

PART TWO
模擬試卷

PART THREE
考生急症室

MOCK PAPER 1: ANSWERS

I. Comprehension (10 questions)

1. E

The answer is "none of these" because the change in time is due to budgetary cutbacks and nothing else.

2. E

'All of us are well aware of the unsound state that the Indian higher education system is in today. Thanks to years of sustained financial neglect, most Indian universities and colleges do no research worth the name'. It is from here we find that the correct answer is option E.

3. B

All these lines can be found easily and that too, word to word. Hence, correct option is option B.

4. C

An easy question. This topic is being talked about in the start and the line that quotes us the answer is '…if one considers the amount of time academics require to prepare lectures of good quality as well as the time they need to spend doing research, it is clear…'.

5. A

Question 5 Explanation:

The comprehension makes very clear that the UGC requires 22 hours of teaching and not 40. Hence option (A) is absolutely false. Rest all other options are correct. Hence, Option (A) is right.

6. E

It a direct statistic which has been asked. The correct option is option (E).

7. D

Option (D) states that public investment is increased in India whereas the passage noticeably states it has been decreased. Hence, the correct option is Option (D).

8. C

'…the UGC says universities must implement a complete freeze on all permanent recruitment…' From the above quote of the passage, it becomes evident that the correct answer is Option (C).

9. C

Both blush and flush mean sudden

reddening of the face (as from embarrassment or guilt or shame or modesty)

10. B

Quantum means a specified portion; a quantity or amount.

II. Error Identification (10 questions)

11. B

- Important file → an important file

12. A

- Having punished → having been punished (passive meaning)

13. C

- protect → protecting

14. A

- More careful → more carefully

15. A

- Know → knows

16. D

- Neither Bob could → Neither could Bob

17. C

- The most → the more

18. B

- Put → to put

19. C

- Safety → safe (adjective)

20. C

- That → which

III. Sentence Completion (10 questions)

21. B

Sentient means possessing the power of sense or sense-perception; conscious.

22. C

To rant means to speak loudly or violently.

23. E

Negligent means to habitually lack in giving proper care or attention; having

PART ONE
題庫練習

PART TWO
模擬試卷

PART THREE
考生急症室

a careless manner.

24. C

Juncture is a point of time, especially one that is at a critical point.

25. E

A martyr is one who sacrifices something of supreme value, such as a life, for a cause or principle; a victim; one who suffers constantly.

26. A

Sibilant means characterized by a hissing sound.

27. B

A rapport is a relationship that is useful and harmonious.

28. D

Manifold means many and varied; of many kinds; multiple.

29. A

Untimely means happening before the proper time.

30. C

Arable means suitable for cultivation, fit for plowing and farming productively.

IV. Paragraph Improvement (10 questions)

31. C

The sentence properly introduces the controversy that is the subject of the passage as one of a number of funny misunderstandings.

Explanation for Incorrect Answer A: Choice (A) is unsatisfactory because it joins two independent thoughts with only a comma.

Explanation for Incorrect Answer B: Choice (B) is unsatisfactory because it is illogical; "it's" does not make sense in this context.

Explanation for Incorrect Answer D: Choice (D) is unsatisfactory because "even" suggests that others do not find the "canali" controversy funny. The passage does not indicate this.

Explanation for Incorrect Answer E: Choice (E) is unsatisfactory because "as funny as" puts the emphasis of the sentence on the other misunderstandings, with the expectation that they will be compared to the "canali" controversy. Such a comparison does not happen in the passage.

32. D

The sentence nicely joins the ideas of sentences 3 and 4 without repeating information unnecessarily.

Explanation for Incorrect Answer A: Choice (A) is unsatisfactory because it is unidiomatic to write that Schiaparelli "studied Mars by" a telescope; "with" or "through" would be more appropriate.

Explanation for Incorrect Answer B: Choice (B) is unsatisfactory because it repeats information about studying/looking at Mars unnecessarily.

Explanation for Incorrect Answer C: Choice (C) is unsatisfactory because it joins two complete thoughts with only a comma.

Explanation for Incorrect Answer E: Choice (E) is unsatisfactory because it unnecessarily separates the act of looking at Mars from the act of studying Mars. As the passage indicates, the activities are the same.

33. E

The word "Because" appropriately signifies the relationship between the translation problem and the misunderstanding about the "canals."

Explanation for Incorrect Answer A: Choice (A) is unsatisfactory because "Whereas" indicates that the mistranslation of "canali" and the misunderstanding about the building of the "canals" were contradictory ideas, when in fact one caused the other.

Explanation for Incorrect Answer B: Choice (B) is unsatisfactory because "However" makes no sense in this context.

Explanation for Incorrect Answer C: Choice (C) is unsatisfactory because "If" suggests that there is some doubt as to whether or not people mistranslated "canali" in this way. The implication of the passage is that people did mistranslate the term, resulting in a misconception about the building of the "canals."

Explanation for Incorrect Answer D: Choice (D) is unsatisfactory because "Although" implies that one would not expect word to spread given that the mistranslation occurred, when the passage indicates the opposite.

34. C

It properly signifies the relationship between the idea presented in the previous sentence and the astronomer Lowell.

Explanation for Incorrect Answer A: Choice (A) is unsatisfactory because the pronoun "them" seems to refer to "intelligent beings" who built canals on Mars, and Lowell was clearly not one of these.

Explanation for Incorrect Answer B: Choice (B) is unsatisfactory because Lowell is not described elsewhere in the passage as intelligent; in fact, his theory is shown to be based on a simple misunderstanding.

Explanation for Incorrect Answer D: Choice (D) is unsatisfactory because it results in an awkward and illogical sentence.

Explanation for Incorrect Answer E: Choice (E) is unsatisfactory because, according to the passage, Lowell created more problems than he solved.

PART ONE
題庫練習
PART TWO
模擬試卷
PART THREE
考生急症室

35. A

The resulting sentence maintains the effective structure of the original sentence 13 while adding the important information from sentence 12.

Explanation for Incorrect Answer B: Choice (B) is unsatisfactory because it is improper to describe a series of books as "a bestseller."

Explanation for Incorrect Answer C: Choice (C) is unsatisfactory because, following sentence 11, it is more appropriate for the subject of the sentence to be "Lowell" than to be "his books."

Explanation for Incorrect Answer D: Choice (D) is unsatisfactory because it does not indicate the relationship between the books Lowell wrote and his popularization of the "intelligent beings" idea.

Explanation for Incorrect Answer E: Choice (E) is unsatisfactory because it suggests that Lowell's books have been previously mentioned in the passage ("In these books"), when this is the first mention of them.

36. D

It properly explains that most children hear the Pocahontas story before they leave elementary school.

Explanation for Incorrect Answer A: Choice (A) is unsatisfactory because the original sentence connects the two main ideas—children leaving school and the Pocahontas story—with only the conjunction "and." The sentence thus offers no clue about the relationship between the two ideas.

Explanation for Incorrect Answer B: Choice (B) is unsatisfactory because it is logical to give the names of the principal figures in a story or event before telling the story, not after—especially when the names are familiar.

Explanation for Incorrect Answer C: Choice (C) is unsatisfactory because it repeats the error of the original in failing to explain the relationship between the children and the story.

Explanation for Incorrect Answer E: Choice (E) is unsatisfactory because the original correctly refers to Pocahontas's tribe.

37. E

The resulting sentence maintains the sense of the original while eliminating the redundancy.

Explanation for Incorrect Answer A: Choice (A) is unsatisfactory because it contains an unclear referent: "which" seems to refer to the incident itself rather than to the report.

Explanation for Incorrect Answer B: Choice (B) is unsatisfactory because the word "which" appears to refer to the incident, when it can logically refer only to the report of the incident.

Explanation for Incorrect Answer C: Choice (C) is unsatisfactory because the word "consequently" suggests incorrectly that Smith's report is a consequence of the legend.

Explanation for Incorrect Answer D: Choice (D) is unsatisfactory because it uses the unnecessary phrase "it seems" to relate a fact.

38. A

It links sentence 10 to the rest of the paragraph by explaining the harmlessness of the "ritual display" mentioned in the previous sentence (and thus clarifies the contrast between Smith's account and the probable facts).

Explanation for Incorrect Answer B: Choice (B) is unsatisfactory because sentence 10 outlines a scenario that challenges Smith's "life-or-death" account, implying that Smith is not a credible source.

Explanation for Incorrect Answer C: Choice (C) is unsatisfactory because the inserted phrase unhelpfully interrupts the connection between the "ritual display" introduced in sentence 9 and the explanation of it in sentence 10.

Explanation for Incorrect Answer D: Choice (D) is unsatisfactory because the use of "and" implies that the "ritual" and the "drama" are two different events, whereas the "drama" actually refers to the "ritual display."

Explanation for Incorrect Answer E: Choice (E) is unsatisfactory because nothing in sentence 10 is contrary to sentence 9; the latter sentence logically follows the former.

39. B

Sentence 10 elaborates on the information about what may have really happened to Smith presented in sentence 9.

Explanation for Incorrect Answer A: Choice (A) is unsatisfactory because sentence 10 offers only support for the claim made in sentence 9.

Explanation for Incorrect Answer C: Choice (C) is unsatisfactory because the information in sentence 10 is not an "example"; rather, it is a reasoned clarification of what may have happened to Smith.

Explanation for Incorrect Answer D: Choice (D) is unsatisfactory because nothing about sentence 10 contradicts sentence 9.

Explanation for Incorrect Answer E: Choice (E) is unsatisfactory because sentence 10 does not make use of any new sources.

40. C

The third paragraph gives two detailed examples of Pocahontas's political successes in later life.

Explanation for Incorrect Answer A: Choice (A) is unsatisfactory because the passage does not mention any of Pocahontas's shortcomings.

Explanation for Incorrect Answer B: Choice (B) is unsatisfactory because focusing on the believability of historical facts is odd and unnecessary.

PART ONE
題庫練習

PART TWO
模擬試卷

PART THREE
考生急症室

Explanation for Incorrect Answer D: Choice (D) is unsatisfactory because the information in paragraph 3 deals primarily with Pocahontas's public life, not her private life.

Explanation for Incorrect Answer E: Choice (E) is unsatisfactory because the third paragraph gives detailed information about Pocahontas that is not in dispute.

Mock Paper 2

Time Limit: 45 minutes

PART ONE
題庫練習

PART TWO
模擬試卷

PART THREE
考生急症室

I. Comprehension (10 questions)

Read the following passage and answer Questions 1-10. For each question, select the best answer from the given alternatives.

At one time it would have been impossible to imagine the integration of different religious thoughts, ideas and ideals. That is because of the closed Society, the lack of any communication or interdependence on other nations. People were happy and content amongst themselves; they did not need anyone any more. The physical distance and cultural barriers prevented any exchange of thought and beliefs. But such is not the case today. Today, the world has become a much smaller place, thanks to the adventures and miracles of science. Foreign nations have become our next-door neighbours. Mingling of population is bringing about an interchange of thought. We are slowly realizing that the world is a single cooperative group. Other religions have become forces with which we have to reckonand we are seeking for ways and means by which we can live together in peace and harmony. We cannot have religious unity and peace so long as we assert that we are in possession of the light and all others are grouping in the darkness. That very assertion is a challenge to a fight. The political ideal of the world is not so much a single empire with a homogeneous, civilization and single communal will a brotherhood of free nations differing profoundlyin life and mind, habits and institutions, existing side by side in peace and order, Harmony and cooperation and each contributing to the world its own unique and specific best, which is irreducible to the terms of the others. The cosmopolitanism of the eighteenth century and the nationalism of the nineteenth are combined in our ideal of a world commonwealth, which allows every branch of the human family to find freedom, security and self-realisation in the larger life of mankind. I see no hope for the religious future of the world, if this ideal is not extended to the religious sphere also. When two or three different systems claim that they contain the revelation of the very core and centre of truth and the acceptance of it is the exclusive pathway to heaven, conflicts are inevitable. In such conflicts one religion will not allow others to steal a march over it and no one can gain ascendancy until the world is reduced to dust and ashes. To obliterate every other religion than one's is a sort of Bolshevism in religion which we must try to prevent. We can do so only if we accept something like the Indian solution, which seeks the unity of religion not in a common creed but in a common quest; Let us believe in a unity of spirit and not of organization, a unity which secures ample liberty not only for every individual but for every type of organized life which has proved itself effective. For almost all historical forms of life and thought can claim the sanction of experience and so the authority of God. The world would be a much poorer thing if one creed absorbed the rest. God wills a rich harmony and not a colourless uniformity. The comprehensive and synthetic spirit of Indianism had made it a mighty forest with a thousand waving

PART ONE
題庫練習

PART TWO
模擬試卷

PART THREE
考生急症室

arms each fulfilling its function and all directed by the spirit of God. Each thing in its place and all associated in the divine concert making with their various voices and even dissonance, as Heracletus would say, the most exquisite harmony should be our ideal.

1. According to the passage. Religious unity and peace can be obtained if-

    A. We believe that the world is a single co-operative group

    B. We do not assert that we alone are in possession of the real knowledge

    C. We believe in a unity of spirit and not of organization.

    D. We believe that truth does not matter and will prevail.

    E. None of these

2. Which of the following, according to the passage, is the Indian solution'? Unity of religions-in a common-

    A. Belief

    B. Organization

    C. Creed

    D. Search

    E. None of these

3. According to the author, which of the following is not true?

   A. Acceptance of Indianism is the exclusive pathway to heaven

   B. We should not assert that other religions have no definite pathway or goal

   C. God wants a genuine similarity in thoughts, ideals and values rather than an artificial appearance

   D. People interacting with each other is bringing about a change in their attitude,

   E. None of these

4. According to the passage, the-political ideal of the contemporary world is to-

   A. Create a single empire with a homogeneous civilization:

   B. Foster the unity of all the religions of the world

   C. Create a world common wealth preserving religious diversity of all the nations

   D. Create brotherhood of free nations who believe in one religion

   E. None of these

PART ONE
題庫練習

PART TWO
模擬試卷

PART THREE
考生急症室

5. According to the passage, the world would be a much poorer thing if-

A. One religion swallows all other religions·

B. One religion accepts the supremacy of other religions

C. Religions adopt toleration as a principle of spiritual life

D. We do not achieve the ideal of brotherhood of free nations

E. None of these

6. Which of the following statements is/are stated or implied in the above passage?

(A) People today are happy and content amongst themselves.

(B) There is no freedom and security in the religious sphere in the world today.

(C) Indianism is directed by the spirit 1 God.

A. Only (A) & (C)

B. Only (B)

C. Only (C)

D. Only (A)

E. None of these

7. According to the passage, what is Bolshevism in the religion?

    A. To ridicule the views sincerely held by others

    B. To accept others religious beliefs and doctrines as authentic as ours

    C. To adhere to rigid dogmatism in religion

    D. To make change in a religion so that it becomes more acceptable

    E. None of these

8. According to the passage, the conflict of religions is inevitable, mainly because each religion-

    A. Believes that anyone who disargrees with it ought to be silenced

    B. Wants to steal a march over others

    C. Claims to possess a complete and exclusive understanding of truth

    D. Believes that the view held strongly by many need not be a correct view

    E. None of these

PART ONE
題庫練習

PART TWO
模擬試卷

PART THREE
考生急症室

9. Identify the correct antonym for the word Superficial from the passage.

   A. Categorization

   B. Profound

   C. Chuckle

   D. Origin

   E. None of these

10. Suggest a suitable replacement for the word in the line. '..as Heracletus would say, the most exquisite harmony should be our ideal.'

   A. Recherché: Lavishly elegant and refined.

   B. Aggregate

   C. Chirpy

   D. Hysterical: State of violent mental agitation.

   E. None of these

II. Error Identification (10 questions)

The sentences below may contain a language error. Identify the part that contains the error of choose "(E) No error" if the entire sentence is correct.

11. My <u>friends and I</u> enjoy playing <u>the guitar</u> <u>and have</u> <u>a walk</u> in the forest.

    A. friends and I

    B. the guitar

    C. and have

    D. a walk

    E. No Error

12. Although their status <u>varies</u> <u>in different countries</u>, <u>but women</u> <u>have gained</u> significant legal rights.

    A. varies

    B. in different countries

    C. but women

    D. have gained

    E. No Error

PART ONE
題庫練習

PART TWO
模擬試卷

PART THREE
考生急症室

13. The books <u>writing</u> <u>by</u> Mark Twain are <u>very</u> popular <u>in</u> the world.

    A. writing

    B. by

    C. very

    D. in

    E. No Error

14. Students <u>can borrow</u> books <u>to read</u>, but they <u>needn't</u> take the books <u>out of the library</u>.

    A. can borrow

    B. to read

    C. needn't

    D. out of the library

    E. No Error

15. We <u>are</u> <u>made</u> <u>learning</u> fifty new words <u>every week</u>.

    A. are

    B. made

    C. learning

    D. every week

    E. No Error

16. He's <u>been studied</u> <u>really hard</u> <u>so that</u> he can <u>pass</u> the exams.

    A. been studied

    B. really hard

    C. so that

    D. pass

    E. No Error

17. No matter <u>what different</u>, <u>various</u> music types have one thing <u>in common</u>: <u>touching</u> the hearts of the listeners.

    A. what different

    B. various

    C. in common

    D. touching

    E. No Error

18. Mary as well as <u>her parents</u> <u>are coming</u> <u>here</u> <u>in</u> a few days.

    A. her parents

    B. are coming

    C. here

    D. in

    E. No Error

PART ONE
題庫練習

PART TWO
模擬試卷

PART THREE
考生急症室

19. The tree <u>has</u> <u>been grown</u> <u>taller than</u> <u>Tim has</u>.

    A. has

    B. been grown

    C. taller than

    D. Tim has

    E. No Error

20. If the <u>little</u> girl knew <u>how to do</u> it, her <u>mother</u> <u>will be</u> pleased.

    A. little

    B. how to do

    C. mother

    D. will be

    E. No Error

III. Sentence Completion (10 questions)

Complete the following sentences by choosing the best answer from the options given.

21. The _____ seventh-grader towered over the other players on his basketball team.

    A. gangling

    B. studious

    C. mimetic

    D. abject

    E. reserved

22. Carson was at first flattered by the _____ of his new colleagues, but he soon realized that their admiration rested chiefly on his connections, not his accomplishments.

    A. reprisal

    B. adulation

    C. bulwark

    D. rapport

    E. retinue

PART ONE
題庫練習

PART TWO
模擬試卷

PART THREE
考生急症室

23. For a(n) _____ fee, it is possible to upgrade from regular gasoline to premium.

    A. nominal

    B. judgmental

    C. existential

    D. bountiful

    E. jovial

24. Searching frantically to find the hidden jewels, the thieves proceeded to _____ the entire house.

    A. justify

    B. darken

    C. amplify

    D. ransack

    E. glorify

25. The _____ deer stuck close to its mother when venturing out into the open field.

    A. starling

    B. foundling

    C. yearling

    D. begrudging

    E. hatchling

26. The police officer _____ the crowd to step back from the fire so that no one would get hurt.

    A. undulated

    B. enjoined

    C. stagnated

    D. permeated

    E. delineated

27. Jackson's poor typing skills were a _____ to finding employment at the nearby office complex.

    A. benefit

    B. hindrance

    C. partiality

    D. temptation

    E. canon

28. Through _____, the chef created a creamy sauce by combining brown sugar, butter, and cinnamon in a pan and cooking them over medium-high heat.

    A. impasse

    B. obscurity

    C. decadence

D. diversion

E. liquefaction

29. The defendant claimed that he was innocent and that his confession was _____.

A. coerced

B. flagrant

C. terse

D. benign

E. futile

30. Harvey was discouraged that his visa application was _____ due to his six convictions.

A. lethargic

B. immeasurable

C. nullified

D. segregated

E. aggravated

IV. Paragraph Improvement

(10 questions)

The sentences below are parts of early drafts of one or more passages, some parts of which may have to be rewritten. Read the draft and choose the best answer to the question.

Passage A

(1) People today have placed emphasis on the kinds of work that others do, it is wrong. (2) Suppose a woman says she is a doctor. (3) Immediately everyone assumes that she is a wonderful person, as if doctors were incapable of doing wrong. (4) However, if you say you're a carpenter or mechanic, some people think that you're not as smart as a doctor or a lawyer. (5) Can't someone just want to do this because he or she loves the work?

(6) Also, who decided that the person who does your taxes is more important than the person who makes sure that your house is warm or that your car runs ? (7) I know firsthand how frustrating it can be. (8) They think of you only in terms of your job. (9) I used to clean houses in the summer because the money was good; but yet all the people whose houses I cleaned seemed to assume that because I

PART ONE
題庫練習

PART TWO
模擬試卷

PART THREE
考生急症室

was vacuuming their carpets I did not deserve their respect. (10) One woman came into the bathroom while I was scrubbing the tub. (11) She kept asking me if I had any questions. (12) Did she want me to ask whether to scrub the tub counter-clockwise instead of clockwise ? (13) Her attitude made me angry! (14) Once I read that the jobs people consider important have changed. (15) Carpenters used to be much more admired than doctors. (16) My point is, then, that who I want to be is much more important than what I want to be!

31. Of the following, which is the best way to phrase sentence 1 reproduced below?

A. (As it is now)

B. People today place too much emphasis on the kinds of work that others do.

C. What kinds of work others do is being placed too much emphasis on by people today.

D. The wrong kind of emphasis had been placed on the kinds of work others do today.

E. The wrong emphasis is being placed today on people and what kind of work they do.

32. In context, which of the following is the best way to revise and combine the underlined portions of sentences 2 and 3?

A. Suppose a woman says she is a doctor, but immediately everyone assumes that she is a wonderful person, as if doctors were incapable of doing wrong.

B. If a woman says she is a doctor, for instance, immediately everyone assumes that she is a wonderful person, as if doctors were incapable of doing wrong.

C. When a woman says she is a doctor, however, immediately everyone assumes that she is a wonderful person, as if doctors were incapable of doing wrong.

D. Immediately, if they say, for example, she is a doctor, everyone assumes that she is a wonderful person, as if doctors were incapable of doing wrong.

E. Therefore, a woman is maybe saying she is a doctor; immediately everyone assumes that she is a wonderful person, as if doctors were incapable of doing wrong.

PART ONE
題庫練習

PART TWO
模擬試卷

PART THREE
考生急症室

33. In context, the phrase do this in sentence 5 would best be replaced by:

    A. hold this particular opinion

    B. resist temptation

    C. ask someone for assistance

    D. become a carpenter or a mechanic

    E. aspire to learn medicine

34. Which of the following is the best way to revise and combine the underlined portions of sentences 7 and 8 reproduced below?

    A. I know firsthand how frustrating it can bbe; they-people, that is-think of you only in terms of your job.

    B. I know firsthand how frustrating it can be when they are thinking of one only in terms of your job.

    C. I know firsthand how frustrating it can be how people think of you only in terms of your job.

    D. I know firsthand how frustrating it can be when people think of you only in terms of your job.

    E. I know firsthand how frustrating it can be; having people think of you only in terms of your job.

35. In context, the phrase but yet in sentence 9 would best be replaced by:

A. incidentally,

B. however,

C. in fact,

D. in addition,

E. for example,

Passage B

(1) Aristotle was a great philosopher and scientist. (2) Aristotle lived in Greece over 2300 years ago. (3) Aristotle was extraordinarily curious about the world around him. (4) He was also a master at figuring out how things worked. (5) Aristotle passed it on to his pupil Theophrastus.

(6) Theophrastus was famous among his contemporaries as the co-founder of the Lyceum, a school in Greece, he is best known today as "the father of botany." (7) Botany is the branch of science dealing with plants.

(8) Two famous books he wrote were Natural History of Plants and Reasons for Vegetable Growth. (9) His books were translated from

PART ONE
題庫練習

PART TWO
模擬試卷

PART THREE
考生急症室

Greek into Latin in 1483—1800 years after he wrote them—they influenced thousands of readers.

(10) Theophrastus made accurate observations about all aspects of plant life, including plant structure, plant diseases, seed use, and medicinal properties. (11) He even described the complex process of plant reproduction correctly, hundreds of years before it was formally proven. (12) In 1694 Rudolph Jakob Camerarius used experiments to show how plants reproduced. (13) According to some accounts, Theophrastus did his research in a garden he maintained at his school which was called the Lyceum. (14) But Theophrastus also wrote about plants that grew only in other countries, which he heard about from returning soldiers. (15) By comparing these plants to plants he grew in his garden, Theophrastus established principles that are still true today.

36. Which of the following is the best version of the underlined portion of sentence 1 and sentence 2 reproduced below?

   A. Aristotle was a great philosopher and a scientist, living in Greece over 2300 years ago.

   B. Aristotle was a great philosopher and scientist who lived in Greece over 2300 years ago.

C. Aristotle was a great philosopher, and, as a scientist, livedin Greece over 2300 years ago.

D. Aristotle was a great philosopher and scientist; Aristotle lived in Greece over 2300 years ago.

E. Aristotle was a great philosopher, scientist, and lived in Greece over 2300 years ago.

37. What would best replace "it" in sentence 5?

A. that

B. them

C. these traits

D. the world

E. his things

38. What word should be inserted between "Greece," and "he" in sentence 6 reproduced below?

A. and

B. but

C. for

D. thus

E. moreover

PART ONE
題庫練習

PART TWO
模擬試卷

PART THREE
考生急症室

39. Which sentence should be inserted between sentence 8 and sentence 9?

A. Theophrastus's ideas had a lasting impact.

B. Theophrastus's books were instantly successful.

C. The first book is still studied today in botany classes.

D. They challenged the conclusions of Aristotle.

E. Theophrastus also taught botany to hundreds of students.

40. Which revision appropriately shortens sentence 13 reproduced below?

A. Delete "his school which was called".

B. Delete "According to some accounts,".

C. Delete "in a garden he maintained".

D. Replace "According to some accounts" with "Therefore".

E. Replace "Theophrastus" with "he".

END OF PAPER

MOCK PAPER 2: ANSWERS

I. Comprehension
(10 questions)

1. C

A very tricky question. The passage has all these options being talked about but only one of them in the context of the question and that option is (C). It can be seen in the second last paragraph.

2. D

The passage says quest. Out of the given options, search is the closest synonym of quest. Hence, the option is (D).

3. A

The confusion arises between option (A) and option (B). '…To obliterate every other religion than one's is a sort of Bolshevism in religion which we must try to prevent..' As option (B) is clearly mentioned, it is the answer. The second best answer is option (A).

4. E

The passage has no discussion regarding the political ideal of the contemporary world. Hence, the answer is none of these.

5. A

The passage mentions something like this '..world would be a much poorer thing if one creed absorbed the rest..' The closest option to this is option (A).

6. B

Nowhere is the mention or implication of 1 God. So, option (C) is excluded. Also, people used to be happy and contented when the societies were closed to each other. So, option (A) is also wrong. Option (B) can be implied from the second paragraph, hence it is the answer.

7. A

Bolshevism was a strategy developed by the Bolsheviks (Russia) between 1903 and 1917 with a view to seizing state power and establishing a dictatorship of the proletariat. They were extremists. Hence after getting to know the meaning the option that can be deduced is option (A).

8. B

All the options make sense but option (B) has the exact words like that of the passage. Hence, we choose option (B).

PART ONE
題庫練習

PART **TWO**
模擬試卷

PART THREE
考生急症室

9. B

Profound Superficial means to be concerned with or comprehending only what is apparent or obvious whereas profound is showing intellectual penetration or emotional depth.

10. A

Exquisite and recherché both mean lavishly elegant and refined.

II. Error Identification
(10 questions)

11. C

- Have a walk → having a walk

12. C

- "but' must be omitted to form the structure with "although"

13. A

- writing must be ed into "written" to form a past participle phrase to say about a passive action 'books written by..."

14. C

- Needn't → mustn't

15. C

- Somebody is made to do something (passive voice) = Make somebody do something (active).

- Learning → to learn

16. A

- Been studied → been studying

17. A

- What different → how different (no matter how + adjective)

18. B

- Are coming → is coming

19. B

- Been grown → grown

20. D

- Conditional sentence of type 2: If subject did / were, ... would do....is used to say about untrue situation at present

- Will be → would be

III. Sentence Completion (10 questions)

21. A

Gangling means awkward, lanky, or unusually tall and thin.

22. B

Adulation means strong or excessive admiration or praise; fawning flattery.

23. A

Nominal means small, virtually nothing, or much below the actual value of a thing.

24. D

To ransack means to thoroughly search, to plunder, pillage.

25. C

A yearling is a young animal past its first year but not yet two years old.

26. B

To enjoin means to issue an order or command; to direct or impose with authority.

27. B

Hindrance is an impediment or obstruction; a state of being hindered; a cause of being prevented or impeded.

28. E

Liquefaction is the process of liquefying a solid or making a liquid.

29. A

To coerce is to force to do through pressure, threats, or intimidation; to compel.

30. C

To nullify means to make invalid or nonexistent.

IV. Paragraph Improvement (10 questions)

31. B

The phrase "too much" smoothly embeds a negative judgment within the first clause, thereby making the second clause unnecessary.

Explanation for Incorrect Answer A: Choice (A) is unsatisfactory because it uses a comma improperly to join two independent clauses.

Explanation for Incorrect Answer C: Choice (C) is unsatisfactory because the word order is garbled. The preposition "on" (one of several misplaced words) is far removed from its object, "kinds."

Explanation for Incorrect Answer D: Choice (D) is unsatisfactory because the past perfect tense ("had been placed") is inappropriate to describe an action occurring today.

Explanation for Incorrect Answer E: Choice (E) is unsatisfactory because of wordiness. The sentence does not need both the noun "people" and the pronoun "they."

32. B

In the combined sentence, the dependent clause (appropriately introduced by "if") states a possible condition, and the main clause then describes a likely result. "For instance" indicates that the situation illustrates the statement in sentence 1.

Explanation for Incorrect Answer A: Choice (A) is unsatisfactory because the connecting word "but" usually introduces a contrast rather than a result.

Explanation for Incorrect Answer C: Choice (C) is unsatisfactory because the transition word "however" inappropriately suggests that a contrasting idea will follow.

Explanation for Incorrect Answer D: Choice (D) is unsatisfactory because it uses a vague pronoun, "they."

Explanation for Incorrect Answer E: Choice (E) is unsatisfactory because the introductory word "therefore" is

inappropriate. It incorrectly suggests that this sentence is drawing a conclusion based on evidence presented earlier.

33. D

In context, the verb "become" is more precise than "do," and the nouns "carpenter" and "mechanic" are much more specific than the pronoun "this."

Explanation for Incorrect Answer A: Choice (A) is unsatisfactory because its implication is illogical. Holding a negative opinion would not be the likely result of loving one's work.

Explanation for Incorrect Answer B: Choice (B) is unsatisfactory because it introduces an irrelevant idea. No other sentence in the passage even mentions temptation.

Explanation for Incorrect Answer C: Choice (C) is unsatisfactory because the relationship between the two clauses becomes illogical. Loving one's work would not necessarily produce a request for assistance.

Explanation for Incorrect Answer E: Choice (E) is unsatisfactory because it shifts emphasis away from the value of work done by carpenters and mechanics. Instead of continuing the idea in sentence 4, this choice introduces a new thought.

34. D

The subordinating conjunction "when" provides an appropriate link between the two clauses, and the noun "people" replaces the vague pronoun "they."

Explanation for Incorrect Answer A: Choice (A) is unsatisfactory because of wordiness. The pronoun "they" and the phrase "that is" are not needed.

Explanation for Incorrect Answer B: Choice (B) is unsatisfactory because it retains the vague pronoun "they" and introduces another inappropriate pronoun, "one" (that is inconsistent with the later pronoun "your").

Explanation for Incorrect Answer C: Choice (C) is unsatisfactory because "how" is not an acceptable transition word to link the first clause (ending with "be") with the second clause.

Explanation for Incorrect Answer E: Choice (E) is unsatisfactory because it uses improper coordination. The semicolon incorrectly links unequal parts (an independent clause before the semicolon and a phrase after it).

35. B

The word "however" properly indicates a contrast between a positive aspect of the job (good pay) and a negative aspect (lack of respect).

Explanation for Incorrect Answer A: Choice (A) is unsatisfactory because "incidentally" does not indicate contrast. Instead, it suggests that the information to follow is of minor importance.

Explanation for Incorrect Answer C:

Choice (C) is unsatisfactory because the phrase "in fact" does not prepare for a contrast. It implies only that the second clause will correct or clarify an earlier misconception.

Explanation for Incorrect Answer D: Choice (D) is unsatisfactory because the phrase "in addition" fails to introduce a contrast. It actually implies that the second clause will continue or reinforce the idea presented earlier.

Explanation for Incorrect Answer E: Choice (E) is unsatisfactory because the phrase "for example" does not imply contrast. It suggests instead that specific details to follow will support an earlier generalization.

36. B

It properly uses a relative clause ("who lived...") to connect the information about Aristotle in sentence 2 to the information in sentence 1.

Explanation for Incorrect Answer A: Choice (A) is unsatisfactory because it results in an awkward sentence.

Explanation for Incorrect Answer C: Choice (C) is unsatisfactory because it unnecessarily separates the idea of Aristotle as a philosopher from the idea of Aristotle as a scientist. According to the passage, he practiced both philosophy and science in Greece 2300 years ago.

Explanation for Incorrect Answer D: Choice (D) is unsatisfactory because the resulting sentence repeats "Aristo-

PART ONE
題庫練習

PART **TWO**
模擬試卷

PART THREE
考生急症室

tle" unnecessarily.

Explanation for Incorrect Answer E: Choice (E) is unsatisfactory because it creates a non-parallel series. If the sentence were "Aristotle was a great philosopher, scientist, and mathematician," such a formulation would be correct, but that is not the case here.

37. C

Sentence 3 and sentence 4 refer to qualities, or traits, that Aristotle possessed. Sentence 5 indicates that Aristotle passed something on to his pupil Theophrastus, and we know from context that Theophrastus possessed both of these traits, so it makes sense to say that Aristotle passed the traits on to Theophrastus.

Explanation for Incorrect Answer A: Choice (A) is unsatisfactory because the paragraph specifies two traits that Aristotle possessed. It is illogical to conclude that he only passed one of them to Theophrastus.

Explanation for Incorrect Answer B: Choice (B) is unsatisfactory because, while Aristotle passed two traits on to Theophrastus, the traits need to be identified as such; "them" is ambiguous in context.

Explanation for Incorrect Answer D: Choice (D) is unsatisfactory because it makes no sense to say that Aristotle passed on "the world" to his student.

Explanation for Incorrect Answer E: Choice (E) is unsatisfactory because,

while Aristotle may have passed "his things" (physical possessions) on to Theophrastus, the passage does not indicate this. On the other hand, the passage makes it clear that Theophrastus possessed both curiosity and ingenuity.

38. B

The conjunction "but" properly indicates the relationship between the two contrasting statements about Theophrastus's fame.

Explanation for Incorrect Answer A: Choice (A) is unsatisfactory because, while "and" creates a correct sentence, it is not as effective in context as "but." The word "and" suggests that the two statements about Theophrastus's fame are complementary facts with no contrast.

Explanation for Incorrect Answer C: Choice (C) is unsatisfactory because the word "for" makes no sense in context.

Explanation for Incorrect Answer D: Choice (D) is unsatisfactory because the word "thus" implies a cause-effect relationship that does not exist.

Explanation for Incorrect Answer E: Choice (E) is unsatisfactory because the word "moreover" signifies that the second statement is "in addition" to the first. "Moreover" does not show the necessary contrast.

39. A

The inserted sentence connects logically to both sentence 8 ("ideas" clearly refers to the content of the books) and sentence 9 (the "lasting impact" is shown by the success of the 1483 translation) and supports the claim made in the first sentence of the paragraph.

Explanation for Incorrect Answer B: Choice (B) is unsatisfactory because, while Theophrastus's books may have been instantly successful (although the fact that he was famous among his contemporaries for other things suggests otherwise), this success is not mentioned in the passage, and is not relevant in context.

Explanation for Incorrect Answer C: Choice (C) is unsatisfactory because the inserted sentence is out of order chronologically. It would make more sense to discuss Theophrastus's relevance today after discussing his relevance in the 1400's.

Explanation for Incorrect Answer D: Choice (D) is unsatisfactory because there is no support in the passage for the claim that Theophrastus's books challenged the conclusions of Aristotle.

Explanation for Incorrect Answer E: Choice (E) is unsatisfactory because any sentence inserted between sentence 8 and sentence 9 should discuss Theophrastus'sbooks (or the ideas in these books), not his teaching.

40. A

The resulting sentence ends "…did his research in a garden he maintained at the Lyceum." As the reader already knows that the Lyceum is the name of Theophrastus's school (sentence 6), this revision is appropriate and necessary.

Explanation for Incorrect Answer B: Choice (B) is unsatisfactory because "According to some accounts" suggests that there is some doubt about the information presented. It would not be appropriate to delete this important qualifier.

Explanation for Incorrect Answer C: Choice (C) is unsatisfactory because it removes the important fact about the garden and maintains the redundancy of the original.

Explanation for Incorrect Answer D: Choice (D) is unsatisfactory because "Therefore" makes no sense in context.

Explanation for Incorrect Answer E: Choice (E) is unsatisfactory because sentence 12 refers to the botanist Camerarius. Sentence 13 must then identify Theophrastus by name, not by "he."

考生急症室一

1） 每隔多久考CRE一次？

CRE一年考兩次，分別在6月和10月考試。

2） 什麼人符合申請資格？

- 持有大學學位（不包括副學士學位）；或

- 現正就讀學士學位課程最後一年；或

- 持有符合申請學位或專業程度公務員職位所需的專業資格。

3） 若然在香港中學文憑考試英國語文科及／或中國語文科取得5級或以上成績，是否需要報考綜合招聘考試英文運用及／或中文運用試卷？

香港中學文憑考試英國語文科5級或以上成績會獲接納為等同綜合招聘考試英文運用試卷的二級成績。香港中學文憑考試中國語文科5級或以上成績會獲接納為等同綜合招聘考試中文運用試卷的二級成績。持有上述成績者不須考試。

PART ONE
題庫練習

PART TWO
模擬試卷

PART THREE
考生急症室

4）若然在香港高級程度會考英語運用科（或General Certificate of Education A Level (GCE A Level) English Language 科）及／或中國語文及文化科取得及格成績，可否獲豁免參加綜合招聘考試？

香港高級程度會考英語運用科或GCE A Level English Language科C級或以上成績會獲接納為等同綜合招聘考試英文運用試卷的二級成績；香港高級程度會考中國語文及文化、中國語言文學或中國語文科C級或以上成績會獲接納為等同綜合招聘考試中文運用試卷的二級成績。如果持有上述成績者不須考試。

香港高級程度會考英語運用科或GCE A Level English Language科D級成績會獲接納為等同綜合招聘考試英文運用試卷的一級成績；香港高級程度會考中國語文及文化、中國語言文學或中國語文科D級成績會獲接納為等同綜合招聘考試中文運用試卷的一級成績。如果持有上述成績，可因應有意投考的公務員職位的要求，決定是否需要應考綜合招聘考試英文運用及／或中文運用試卷。

5）「綜合招聘考試」(CRE)跟「聯合招聘考試」(JRE)有何分別？

在CRE中英文運用考試中取得「二級」成績後，可投考JRE，考試為AO、EO及勞工事務主任、貿易主任四職系的招聘而設。

6) CRE成績何時公佈？

考試邀請信會於考前12天以電郵通知，成績會在試後1個月內郵寄到考生地址。

7) 報考CRE的費用是多少？

不設收費。

8) 若然在綜合招聘考試的英文運用及中文運用試卷取得二級或一級成績，並在能力傾向測試中取得及格成績，是否已符合資格申請公務員職位？可以在何時及怎樣申請這些職位？

個別進行招聘的部門/職系會於招聘廣告中列明有關職位所需的綜合招聘考試成績。由於綜合招聘考試與公務員職位的招聘程序是分開進行的，應留意在各報章及公務員事務局網頁刊登的公務員職位招聘廣告，然後直接向進行招聘的部門/職系提交職位申請。進行招聘的部門/職系會核實你的學歷及/或專業資格，並可能在綜合招聘考試外，另設其他考試/面試。

PART ONE
題庫練習

PART TWO
模擬試卷

PART THREE
考生急症室

9) 可否使用CRE的成績來申請政府以外的工作？

CRE招聘考試是為招聘學位或專業程度公務員職位而設的基本測試，而非一項學歷資格。

10) 如遺失了CRE考試／基本法測試的成績通知書，可否申請補領？

可以書面（地址：香港添馬添美道2號政府總部西翼7樓718室）或電郵形式（電郵地址：csbcseu@csb.gov.hk）向公務員考試組提出申請。

看得喜 放不低

創出喜閱新思維

書名	公務員招聘 英文運用 CRE 解題王
ISBN	978-988-76629-0-7
定價	HK$138
出版日期	2023年9月
作者	FONG SIR
責任編輯	鄭浩文
版面設計	吳芷菁
出版	文化會社有限公司
電郵	editor@culturecross.com
網址	www.culturecross.com
發行	聯合新零售（香港）有限公司
	地址：香港鰂魚涌英皇道1065號東達中心1304-06室
	電話：（852）2693 5300
	傳真：（852）2565 0919

網上購買 請登入以下網址：

一本 My Book One　　　　香港書城 Hong Kong Book City

🌐 (www.mybookone.com.hk)　🌐 (www.hkbookcity.com)